Linear
Associative
Algebras

Linear Associative Algebras

Alexander Abian
Professor of Mathematics
Iowa State University

PERGAMON PRESS INC.
New York   Toronto   Oxford   Sydney   Braunschweig

Pergamon Press Inc., Maxwell House, Fairview Park, Elmsford, N.Y. 10523
Pergamon of Canada Ltd., 207 Queen's Quay West, Toronto 117, Ontario
Pergamon Press Ltd., Headington Hill Hall, Oxford
Pergamon Press (Aust.) Pty. Ltd., Rushcutters Bay, Sydney, N.S.W.
Vieweg & Sohn GmbH, Burgplatz 1, Braunschweig

Printed in the United States of America

08 016564 8(H)

Contents

Preface

This book is a self-contained exposition of Finite Dimensional Linear Associative Algebras culminating with the Wedderburn structure theorems. The text can be used in a graduate Algebra course and can be covered within one semester.

Since a Linear Associative Algebra has the properties of a ring as well as a vector space, the first chapter of the book reviews some preliminary material in connection with rings and the second chapter deals with vector spaces. We assume that the reader has some acquaintance with this material. Chapters three and four treat the essential topics of the subject matter. Because of the importance of matrices in Linear Associative Algebras, the theory of matrices is developed to a sufficient degree and parallel with the development of the text.

The theory of Finite Dimensional Linear Associative Algebras is a special instance of the theory of Rings with Minimum Condition. Nevertheless, the independent study of the Finite Dimensional Linear Associative Algebras is of fundamental significance from both mathematical and pedagogical points of view.

It is my intention to keep the exposition of the material as lucid as possible. There are many motivating and explanatory passages and examples in the text. Also, there is a list of problems at the end of each section.

The author thanks Dr. Jules Borenstein for helpful suggestions.

Iowa State University   ALEXANDER ABIAN
Ames, Iowa

CHAPTER 1

Preliminaries

1.1. Semigroups and Groups

The concept of a *binary operation* on a set is so fundamental in Algebra that we may say an algebraic system is created as soon as one or more binary operations are defined on a set.

By a binary operation on a set $\mathscr{A}$ we mean a mapping, say, $*$ from the cartesian product $\mathscr{A} \times \mathscr{A}$ into $\mathscr{A}$. Thus, for every ordered pair (X,Y) of elements X and Y of $\mathscr{A}$ the binary operation $*$ assigns a *unique* mate Z which is an element of $\mathscr{A}$ and which is denoted by $*(X,Y)$ or, more customarily, by $X * Y$. To indicate that Z and $X * Y$ stand for the same element of $\mathscr{A}$ we write, as usual:

$$X * Y = Z$$

where the *equality* $=$ is taken in the sense of the set-theoretical identity with its usual properties.

If $*$ is a binary operation on the set $\mathscr{A}$ then the ordered pair $(\mathscr{A}, *)$ is called the *algebraic system* defined by $*$.

In order to define an algebraic system $(\mathscr{A}, *)$ it is sufficient to specify the set $\mathscr{A}$ and the unique mate $X * Y$ corresponding to every element X and Y of $\mathscr{A}$. The specification of $X * Y$ is usually accomplished in a variety of ways: by operational tables, by formulas, by ordinary sentences and so forth.

If $(\mathscr{A}, *)$ is an algebraic system, for the sake of simplicity, we often refer to the set $\mathscr{A}$ as the algebraic system under consideration and we refer to the elements of $\mathscr{A}$ as the elements of $(\mathscr{A}, *)$.

Similarly, if $*$ and $\odot$ are binary operations on the set $\mathscr{A}$ then the ordered triple $(\mathscr{A}, *, \odot)$ is called the algebraic system defined by $*$ and $\odot$.

Again, in order to define an algebraic system $(\mathscr{A}, *, \odot)$ it is sufficient to specify the set $\mathscr{A}$ and the unique mates $X * Y$ and $X \odot Y$ corresponding to every element X and Y of $\mathscr{A}$.

Here again, if $(\mathscr{A}, *, \odot)$ is an algebraic system, for the sake of simplicity, we refer to the set $\mathscr{A}$ as the algebraic system under consideration and we refer to the elements of $\mathscr{A}$ as the elements of $(\mathscr{A}, *, \odot)$.

The development of an algebraic system consists of deriving necessary conclusions from those hypotheses that are imposed on the binary operations which define the algebraic system. These hypotheses are called the *axioms* of the algebraic system.

From the above it is clear that in order to develop an algebraic system a necessary amount of knowledge of mathematical logic and Set Theory is required. We shall assume this knowledge on the part of the reader.

A reasonably sufficiently developed algebraic system is usually called an *algebraic theory*. In this book we are concerned with the exposition of that algebraic theory which is called Finite Dimensional Linear Associative Algebra.

Below we mention some hypotheses of a general nature each of which may or may not be satisfied by a binary operation $*$ on a set $\mathscr{A}$:

Associativity of $*$, *i.e.*,

$$(X * Y) * Z = X * (Y * Z) \tag{1}$$

for every element X, Y and Z of $\mathscr{A}$.

Commutativity of $*$, *i.e.*,

$$X * Y = Y * X \tag{2}$$

for every element X and Y of $\mathscr{A}$.

Solvability of equations; *i.e.*, for every element A and B of $\mathscr{A}$ each of the equations

$$A * X = B \quad \text{and} \quad Y * A = B \tag{3}$$

has a solution in $\mathscr{A}$.

Existence of a neutral element, *i.e.*, there exists an element (which is easily seen to be unique) N of $\mathscr{A}$ such that

$$N * X = X * N = X \tag{4}$$

for every element X of $\mathscr{A}$.

Existence of inverses of elements, *i.e.*, for every element X of $\mathscr{A}$ there exists an element Y of $\mathscr{A}$, called an inverse of X, such that:

$$Y * X = X * Y = N \tag{5}$$

where N is the neutral element of $\mathscr{A}$.

Cancellation laws, *i.e.*, each of the equations

$$X * A = X * B \quad \text{and} \quad A * X = B * X \tag{6}$$

implies $A = B$ for every element X, A and B of $\mathscr{A}$.

Some additional hypotheses which modify the ones mentioned above are also often considered. These additional hypotheses give rise to familiar notions such as: *left neutral* element, *right neutral* element, *left inverse* of an element, *right inverse* of an element, *left cancellation* law, *right cancellation* law.

In connection with an algebraic system $(\mathscr{A}, *, \odot)$ the following hypothesis of general nature also may or may not be satisfied by the binary operations $*$ and $\odot$.

Distributivity of, say, $\odot$ with respect to $*$, *i.e.*,

$$X \odot (Y * Z) = (X \odot Y) * (X \odot Z)$$

and

$$(Y * Z) \odot X = (Y \odot X) * (Z \odot X) \tag{7}$$

for every element X, Y and Z of $\mathscr{A}$.

Here again the hypothesis of distributivity may be refined to *left distributivity* and *right distributivity*.

If $*$ is an associative (or commutative) binary operation on the set $\mathscr{A}$ then, as expected, $(\mathscr{A}, *)$ or simply $\mathscr{A}$ is called an *associative* (or a *commutative*) *algebraic system*.

Let us mention some of the consequences of the associativity and commutativity of a binary operation.

Let $(\mathscr{A}, *)$ be an algebraic system. Since $*$ is a binary operation the expression $X * Y * Z$ for elements X, Y and Z of $\mathscr{A}$ is meaningless unless it is parenthesized as $(X * Y) * Z$ or $X * (Y * Z)$. Clearly, in general, these two need not be equal. However, if $*$ is associative then in view of (1) we may let (as we shall do) the expression $X * Y * Z$ designate the unique element of $\mathscr{A}$ which is equal to either $(X * Y) * Z$ or $X * (Y * Z)$.

From the above it follows by induction that if $(\mathscr{A}, *)$ is an associative algebraic system we may let $X_1 * X_2 * \cdots * X_n$ designate the unique element of $\mathscr{A}$ which is obtained by parenthesizing $X_1 * X_2 * \cdots * X_n$ in any legitimate way. If $*$ in addition to being associative is also commutative then in view of (2) we see that $X_1 * X_2 * \cdots * X_n$ remains unchanged no matter how the elements $X_1, X_2, \ldots, X_n$ are permuted among each other.

Again, let $(\mathscr{A}, *)$ be an associative algebraic system. Then if an element X of $\mathscr{A}$ has an inverse it is necessarily unique. For let Y and Y' each be an inverse of X. Then by (1), (4) and (5), we have:

$$Y = (Y' * X) * Y = Y' * (X * Y) = Y'.$$

Thus, $Y = Y'$ and X has a unique inverse.

If the symbol $+$ is used to denote a binary operation on a set $\mathscr{A}$ then $+$ is called *addition* and $(\mathscr{A}, +)$ is called an *additive system*. In this case $X + Y$ is called the *sum* of the *summands* X and Y (in this order) for every element X and Y of $\mathscr{A}$.

If the neutral element of $(\mathscr{A}, +)$ exists then it is usually denoted by 0 and is called the *zero* element of $\mathscr{A}$. If an element X of $(\mathscr{A}, +)$ has a unique inverse then this inverse is usually denoted by $-X$ and is called the *additive inverse* of X. Clearly,

$$(-X) + X = X + (-X) = 0.$$

Moreover, if Y is any element of $(\mathscr{A}, +)$ and $-X$ is the unique inverse of X then $Y + (-X)$ is usually denoted by $Y - X$ and is called the *difference* of Y and X (in this order).

If $(\mathscr{A}, +)$ is an associative additive system then for every positive natural number N the sum of n summands each equal to X is denoted by nX when no confusion is likely to arise.

If the symbol $\cdot$ is used to denote a binary operation on a set $\mathscr{A}$ then $\cdot$ is called *multiplication* and $(\mathscr{A}, \cdot)$ is called a *multiplicative system*. In this case $X \cdot Y$ is most often denoted simply by XY and is called the *product* of the *factors* X and Y (in this order).

If the neutral element of $(\mathscr{A}, \cdot)$ exists then it is usually denoted by I and is called the unit element or *unity* of $\mathscr{A}$. If an element of $(\mathscr{A}, \cdot)$ has a unique inverse then it is usually denoted by X^{-1} and is called the *multiplicative inverse* of X. Clearly,

$$X^{-1}X = XX^{-1} = I.$$

If $(\mathscr{A}, \cdot)$ is an associative multiplicative system then for every positive natural number n the product of n factors each equal to X is denoted by X^n.

Let $(\mathscr{A}, *)$ and $(\mathscr{A}', *')$ be two algebraic systems. Clearly, from the algebraic point of view they are indistinguishable if there is a one-to-one correspondence φ between $\mathscr{A}$ and $\mathscr{A}'$ which also preserves the operations, *i.e.*,

$$\varphi(X * Y) = \varphi(X) *' \varphi(Y) \tag{8}$$

for every element X and Y of $\mathscr{A}$. Such a one-to-one correspondence is called an *isomorphism* and the corresponding systems are called *isomorphic*. Clearly, isomorphism is an equivalence relation in any set of algebraic systems.

If $(\mathscr{A}, *)$ and $(\mathscr{A}', *')$ are isomrophic algebraic systems then we denote this by $(\mathscr{A}, *) \cong (\mathscr{A}', *')$ or simply by $\mathscr{A} \cong \mathscr{A}'$. If $\mathscr{A} \cong \mathscr{A}'$ then we shall often say that $\mathscr{A}$ can be *represented* by $\mathscr{A}'$ and vice versa; or, that $\mathscr{A}'$ *represents* $\mathscr{A}$ and vice versa.

It is obvious how the notion of isomorphism is extended to the case of algebraic systems each with more than one binary operation.

The notion of isomorphism can be weakened to that of *homomorphism*. A mapping φ is called a *homomorphism* from $(\mathscr{A}, *)$ into $(\mathscr{A}', *')$ if and only if φ is a mapping from $\mathscr{A}$ into $\mathscr{A}'$ satisfying (8).

Let $*$ be a binary operation on a non-empty set $\mathscr{A}$. If no additional hypotheses are postulated then $(\mathscr{A}, *)$ is called a *groupoid*.

Examples of groupoids are numerous. For instance, if $\mathscr{P} = \{A, B\}$ and $*$ is defined by

$$A * A = A * B = B * B = B \qquad \text{and} \qquad B * A = A$$

then $(\mathscr{P}, *)$ is a groupoid. Clearly, $(\mathscr{P}, *)$ is neither associative nor commutative, nor does it have a neutral element. If it is postulated that a binary operation $*$ on a non-empty set $\mathscr{A}$ is associative then $(\mathscr{A}, *)$ is called a *semigroup*.

For instance, if $\mathscr{S} = \{C, D\}$ and $*$ is defined by:

$$C * C = C * D = C \qquad \text{and} \qquad D * C = D * D = D$$

then $(\mathscr{S}, *)$ is a semigroup. Clearly, $(\mathscr{S}, *)$ is neither commutative nor does it have a neutral element.

Thus, a semigroup is an associative groupoid. In this connection we mention only that if it is postulated that in a groupoid each of the equations mentioned in (3) has a unique solution then the resulting system is called a *quasigroup*. Moreover, if a quasigroup has a neutral element then it is called a *loop*.

Familiar examples of semigroups are numerous. For instance, the set of all n by n matrices with natural number entries is a semigroup with respect to the usual matrix multiplication.

If $*$ is a binary operation on a nonempty set $\mathscr{A}$ and if it is postulated that $*$ is associative and that each of the equations mentioned in (3) has a solution in $\mathscr{A}$ then $(\mathscr{A}, *)$ is called a *group*.

It is readily proved that if $*$ is an associative binary operation on a nonempty set $\mathscr{A}$ and if each of the equations mentioned in (3) has a solution in $(\mathscr{A}, *)$ then these solutions are unique. Moreover, in a group the cancellation laws (6) hold. Furthermore, $(\mathscr{A}, *)$ has a neutral element and each element of $\mathscr{A}$ has a unique inverse.

Thus, a group has a neutral element and each element of a group has a unique inverse. Moreover, it is easy to see that an associative quasigroup as well as an associative loop is a group.

Familiar examples of groups are numerous. For instance, the set of all nonsingular n by n matrices with rational, real, complex or real quaternion number entries is a group with respect to the usual matrix multiplication.

If $(\mathscr{A}, *)$ is a group and if it is postulated that $*$ is commutative then $(\mathscr{A}, *)$ is called a *commutative* or *abelian* group.

For instance, if $\mathscr{G} = \{E\}$ then $(\mathscr{G}, *)$ is an abelian group.

We assume that the reader has some knowledge of the notions of subgroup, normal subgroup, quotient group and various standard homomorphism and isomorphism theorems involving groups.

Exercises

1. Consider the set ω of all natural numbers $0, 1, 2, \ldots$ and the binary operation $*$ on ω defined by $X * Y = X$ for every element X and Y of ω. Prove that $(\omega, *)$ is a semigroup and that every element of ω is a right neutral element of $(\omega, *)$.

2. Consider the set ω of all natural numbers and the binary operation $*$ on ω defined by $X * Y = X + Y + XY$. Is $(\omega, *)$ a semigroup? Does $(\omega, *)$ have a left or right neutral element?

3. Prove that there exists a denumerable set of semigroups each with a neutral element such that every finite semigroup with a neutral element is isomorphic to one of these semigroups.

4. Let φ be a homomorphism from a group $\mathscr{A}$ onto a group $\mathscr{A}'$. Prove that φ is an isomorphism if and only if $\varphi(X) = N'$ implies $X = N$ for every element X of $\mathscr{A}$ where N is the neutral element of $\mathscr{A}$ and N' that of $\mathscr{A}'$.

5. Let $(\mathscr{A}, +)$ be a group. Prove that the mapping φ from the set of all integers into $\mathscr{A}$ is a homomorphism if $\varphi(n) = nA$ where A is an element of $\mathscr{A}$.

6. Let $(\mathscr{A}, \cdot)$ be a group and A an element of $\mathscr{A}$. Prove that the mapping φ from $\mathscr{A}$ onto $\mathscr{A}$ with $\varphi(X) = AXA^{-1}$ is an automorphism of $\mathscr{A}$ (*i.e.*, an isomorphism from $\mathscr{A}$ onto $\mathscr{A}$).

7. Prove that a semigroup is a group if it has a left (or right) neutral element with respect to which every element of the semigroup has a left (or right) inverse.

8. Prove that a finite semigroup is a group if the cancellation laws hold in the semigroup.

9. Prove that if the cancellation laws hold in a semigroup then it is isomorphic to a subsemigroup of a group.

10. Prove that a commutative semigroup is isomorphic to a subsemigroup of an abelian group if and only if the cancellation laws hold in the semigroup.

11. Prove that a group with less than six elements is an abelian group. Prove also that there are only two non-isomorphic groups each with six elements and that there are only five pairwise non-isomorphic groups each with eight elements.

12. Prove that a group $(\mathscr{A}, \cdot)$ is commutative if and only if the mapping φ from $\mathscr{A}$ onto $\mathscr{A}$ with $\varphi(X) = X^{-1}$ is an isomorphism.

13. Let $(\mathscr{A}, +)$ be a group with 15 elements. Prove that $\mathscr{A}$ has an element A such that every element of $\mathscr{A}$ is equal to nA for some natural number n.

14. Prove that if a group has an even number of elements then it has an even number of elements each of which is its own inverse.

15. Let $\mathscr{B}$ and $\mathscr{C}$ be two distinct subgroups of a group $\mathscr{A}$ each containing at least two elements. Prove that $\mathscr{A}$ contains an element which is contained in neither $\mathscr{B}$ nor $\mathscr{C}$.

16. Using (7) define self-distributivity of a binary operation. Prove that a group is not self-distributive in general.

17. Let $(\mathscr{A}, \cdot)$ be a group. Prove that if X and Y are elements of $\mathscr{A}$ then the set of all the products of the form $XYX^{-1}Y^{-1}$ (called the commutator of X and Y) forms a normal subgroup (called the commutator subgroup of $\mathscr{A}$).

18. Prove that every subquasigroup of a loop is a subloop of that loop.

19. Prove that a loop with less than five elements is a group.

20. Give an example of a loop with five elements which is not a group.

1.2. Rings and Fields

We next consider algebraic systems with two binary operations.

Let $\mathscr{R}$ be a nonempty set and $+$ (addition) and $\cdot$ (multiplication) be two binary operations on $\mathscr{R}$ such that the following three conditions are satisfied:

$(\mathscr{R}, +)$ is an abelian group,
$(\mathscr{R}, \cdot)$ is a semigroup,
$\cdot$ is distributive with respect to $+$.

Then $(\mathscr{R}, +, \cdot)$, or simply, $\mathscr{R}$ is called a *ring*.

Let $(\mathscr{R}, +, \cdot)$ be a ring. We call $(\mathscr{R}, +)$ *the additive part* and $(\mathscr{R}, \cdot)$ *the multiplicative part* of the ring.

As expected, the zero of the additive part of a ring is called the *zero of the ring* and is denoted by 0, as mentioned on page 4. Moreover, the additive inverse of an element X of $\mathscr{R}$ is denoted by $-X$, as mentioned on page 4. Also, it is readily seen that for every element X and Y of $\mathscr{R}$

$$0X = X0 = 0, \qquad (-X)(-Y) = XY$$

and

$$(-X)(Y) = X(-Y) = -(XY)$$

where in the above products $\cdot$ is dropped, as mentioned on page 4. For the sake of simplicity $-(XY)$ will be denoted by $-XY$. Furthermore, for every element X, Y and Z of $\mathscr{R}$ the following distributive laws hold:

$$X(Y \pm Z) = XY \pm XZ \qquad \text{and} \qquad (Y \pm Z)X = YX \pm ZX.$$

Using the notation introduced on page 4, it can be easily verified that in a ring the following *exponentiation* laws hold:

$$X^m X^n = X^{m+n} \quad \text{and} \quad (X^m)^n = X^{mn} \tag{9}$$

for every element X of the ring and every positive natural number m and n.

Familiar examples ot rings are numerous. For instance, if $(\mathcal{R}, +)$ is an abelian group and if $(\mathcal{R}, \cdot)$ is defined by

$$XY = 0$$

for every element X and Y of $\mathcal{R}$ then $(\mathcal{R}, +, \cdot)$ is a ring. Such a ring is called a *zero ring*.

If the multiplicative part of a ring is a commutative semigroup then the ring is called *commutative*.

In a commutative ring, in addition to the above exponentiation laws, we have

$$(XY)^m = X^m Y^m \tag{10}$$

for every element X and Y of the ring and every positive natural number m.

If the multiplicative part of a ring has a unit element then it is called the *unit element*, or simply the *unity* of the ring and is denoted by I, as mentioned on page 4.

A ring which has a unit element is called a *ring with unity*.

We note that in a ring $0 = I$ if and only if the ring has precisely one element.

Examples of commutative rings with unity are numerous. For instance, each of the sets of all integers, rational, real and complex numbers is a commutative ring with unity under the usual addition and multiplication. So is the set $\{0, 1, 2, \ldots, n\}$ with addition and multiplication performed modulo $n + 1$.

On the other hand, the ring of all even integers is a commutative ring without unity (under the usual addition and multiplication).

The set of all n by n matrices with $n > 1$ whose entries are integers (or rational, or real, or complex, or quaternion numbers) is a non-commutative ring with unity under the usual matrix addition and multiplication.

On the other hand, the set of all n by n matrices with $n > 1$ whose entries are even integers is a non-commutative ring without unity under the usual matrix addition and multiplication.

In a ring with unity if the multiplicative inverse of an element X exists then it is denoted by X^{-1} as mentioned on page 4.

An element of a ring which has a multiplicative inverse is called a *nonsingular* element, otherwise, it is called a *singular* element of the ring. Clearly, the set of all nonsingular elements of a ring is a group under the ring multiplication.

If X and Y are nonsingular elements of a ring then

$$(XY)^{-1} = Y^{-1}X^{-1}.$$

Moreover, for every element X of a ring we define

$$X^0 = I = 0^0$$

where 0 is the zero natural number. Clearly, $(X^{-1})^m = (X^m)^{-1}$ where m is any positive natural number. Then for every integer m and n the exponentiation laws (9) hold in any ring and (9) and (10) hold in any commutative ring.

A nonzero element A of a ring is called a *divisor of zero* if

$$AB - 0 \qquad \text{or} \qquad CA - 0$$

for some nonzero element B or C of the ring. Clearly, if A is a divisor of zero then A is singular.

An element N of a ring is called *nilpotent of index k* if

$$N^k = 0 \qquad \text{and} \qquad N^{k-1} \neq 0$$

for some positive natural number k. The zero element of a ring is the only nilpotent element of index 1 of the ring.

If N is a nilpotent element of index $k \geq 2$ then N is a divisor of zero since $NN^{k-1} = 0$.

An element E of a ring is called *idempotent* if

$$E^2 = E.$$

Boolean rings provide examples of rings in which every element is idempotent.

If there exists the smallest positive natural number n such that (using notation introduced on page 4) $nX = 0$ for every element X of a ring and $(n-1)Y \neq 0$ for some element Y of the ring then the ring is said to have *characteristic n*. Otherwise, the ring is said to have *characteristic zero*.

The set $\{0, 1, 2, 3, 4, 5\}$ with the usual addition and multiplication modulo 6 is an example of a commutative ring with unity, with divisors of zero and with characteristic equal to 6.

A commutative ring with unity, without divisors of zero and having more than one element, is called an *integral domain*.

For instance, the set of all integers with the usual addition and multiplication is an integral domain. So is the set $\{0, 1, 2, 3, 4\}$ with addition and multiplication performed modulo 5.

9

The characteristic of an integral domain is easily seen to be either a prime number or zero.

A ring whose nonzero elements form a group under the ring multiplication is called a *division ring*.

Thus, a division ring $(\mathcal{D}, +, \cdot)$ has more than one element, has no divisors of zero, has either a prime or zero characteristic, has a unity and every nonzero element of $\mathcal{D}$ is nonsingular.

Clearly, in a division ring $(\mathcal{D}, +, \cdot)$ for every element B of $\mathcal{D}$ and every nonsingular element A of $\mathcal{D}$ each of the equations

$$AX = B \qquad \text{and} \qquad XA = B$$

has a unique solution, respectively equal to $A^{-1}B$ and BA^{-1}.

Familiar examples of division rings are numerous. For instance, the set of all real quaternions is a division ring under their usual addition and multiplication.

A commutative division ring is called a *field*.

It is readily seen that every finite integral domain is a field. Moreover, every finite division ring is a field. Thus, all finite division rings are *Galois fields* each having p^m elements for some prime p and some positive natural number m.

The set of all rational, real and complex numbers are familiar examples of fields with their usual addition and multiplication.

Let us observe that in a field $\mathcal{F}$ a polynomial of first degree in X such as

$$AX + C$$

with $A \neq 0$ and C elements of $\mathcal{F}$ has a *root*, *i.e.*, there exists an element R of $\mathcal{F}$ such that $AR + C = 0$.

Clearly, not every polynomial (of degree > 1) in X whose coefficients are elements of a field has a root in that field. A field in which every non-constant polynomial (whose coefficients are elements of that field) has a root in that field is called an *algebraically* closed field. For instance, the field of complex numbers is an algebraically closed field.

Using Zorn's lemma, it can be readily proved that every field is a subfield of an algebraically closed field.

The reader should be acquainted with the notions of subring, ideal and quotient ring, as well as various standard homomorphism and isomorphism theorems involving rings. Below, we give a brief description of these notions.

A non-empty subset of a ring is called a *subring* if the subset itself is a ring with respect to the addition and multiplication of the ring.

For instance, the set of all elements of a ring $\mathcal{R}$ which commute with every element of $\mathcal{R}$ is a commutative subring of $\mathcal{R}$ and is called the *center* of $\mathcal{R}$.

A subring $\mathscr{I}$ of a ring $\mathscr{R}$ is called a *two sided ideal*, or simply, an *ideal* of $\mathscr{R}$ if $\mathscr{I}$ is closed with respect to the multiplication by the elements of $\mathscr{R}$, i.e.,

$$RA \in \mathscr{I} \quad \text{and} \quad AR \in \mathscr{I}$$

for every element A of $\mathscr{I}$ and every element R of $\mathscr{R}$.

If the first of the above mentioned membership relations holds then $\mathscr{I}$ is called a *left ideal* of $\mathscr{R}$ and if the second holds then $\mathscr{I}$ is called a *right ideal* of $\mathscr{R}$. These ideals are called *one-sided* ideals of $\mathscr{R}$.

Clearly, in a ring $\mathscr{R}$ the singleton $\{0\}$ and $\mathscr{R}$ itself are ideals of $\mathscr{R}$. These two ideals are called the *trivial ideals* of $\mathscr{R}$ and any other ideal of $\mathscr{R}$ is called *nontrivial*.

If a ring $\mathscr{R}'$ is a homomorphic image of a ring $\mathscr{R}$ then the set of all elements of $\mathscr{R}$ whose image is the zero of $\mathscr{R}'$ is an ideal of $\mathscr{R}$ which is called the *kernel* of the homomorphism under consideration.

Let A be an element of a ring $\mathscr{R}$. Then the set of all elements XA (or AX) with $X \in \mathscr{R}$ is a left (or right) ideal of $\mathscr{R}$ which may not contain A unless $\mathscr{R}$ has a unity. Moreover, the set of all elements of $\mathscr{R}$, each of which is a finite sum of summands each of the form XAY with X and Y elements of $\mathscr{R}$, is an ideal of $\mathscr{R}$ which may not contain A unless $\mathscr{R}$ has a unity.

The intersection of all (left, right) ideals of a ring $\mathscr{R}$, each containing a non-empty subset $\mathscr{E}$ of $\mathscr{R}$, is called the (left, right) *ideal generated* by $\mathscr{E}$. If A is an element of $\mathscr{R}$ then the (left, right) ideal generated by $\{A\}$ is simply called a (left, right) *principal ideal* of $\mathscr{R}$ *generated by A*.

There are some rings every ideal of which is principal. An example of such a ring is the ring of all integers with their usual addition and multiplication.

A ring which is not a zero ring and which has no nontrivial ideal is called a *simple ring*. Some authors call a ring *simple* if it has no nontrivial ideal. However, for our purpose, our definition is more convenient.

It can be readily shown that a ring is a division ring if and only if it is a nonzero ring with no left or right nontrivial ideal.

An (left, right) ideal $\mathscr{M}$ of a ring $\mathscr{R}$ is called a (left, right) *maximal* ideal of $\mathscr{R}$ if $\mathscr{M} \neq \mathscr{R}$ and $\mathscr{M} \subset \mathscr{I}$ implies $\mathscr{M} = \mathscr{I}$ or $\mathscr{R} = \mathscr{I}$ for every (left, right) ideal $\mathscr{I}$ of $\mathscr{R}$.

Clearly, a *quotient ring* $\mathscr{R}/\mathscr{M}$ is a simple ring if and only if $\mathscr{M}$ is a maximal ideal of the ring and $AB \notin \mathscr{M}$ for some element A and B of $\mathscr{R}$. Moreover, if $\mathscr{R}$ is a commutative ring with unity then the quotient ring $\mathscr{R}/\mathscr{M}$ is a field if and only if $\mathscr{M}$ is a maximal ideal of $\mathscr{R}$.

An ideal $\mathscr{P}$ of a ring $\mathscr{R}$ is called a *prime* ideal of $\mathscr{R}$ if $AB \in \mathscr{P}$ implies $A \in \mathscr{P}$ or $B \in \mathscr{P}$ for every element A and B of $\mathscr{R}$.

Clearly, a quotient ring $\mathscr{R}/\mathscr{P}$ is a ring without divisors of zero if and only if $\mathscr{P}$ is a prime ideal of the ring $\mathscr{R}$. Moreover, if $\mathscr{R}$ is a commutative ring with unity then $\mathscr{R}/\mathscr{P}$ is an integral domain if and only if $\mathscr{P}$ is a prime ideal of $\mathscr{R}$.

Let $\mathscr{R}$ be a ring without unity. Then there exists a ring $\mathscr{R}'$ with unity and with the same characteristic as that of $\mathscr{R}$ such that $\mathscr{R}$ is an ideal of $\mathscr{R}'$. In short, a ring without unity can be embedded (as an ideal) in a ring with unity. In this connection, let us also mention that an integral domain can always be embedded (as a subring) in a field.

In our definition of a ring the associativity of the multiplication was postulated. If in the definition of a ring the associativity of the multiplication is denied then the resulting algebraic system is called a *nonassociative ring*.

Important examples of algebraic systems which in general are non-associative rings are the alternative rings, power-associative rings, Lie rings and Jordan rings.

Let us observe that the axiom of associativity of the multiplication of a ring $\mathscr{R}$ is equivalent to stipulating that every subring of $\mathscr{R}$ which is generated by three elements is associative.

Motivated by the above, we define alternative and power-associative rings as follows.

In what follows let $(\mathscr{T}, +, \cdot)$ denote an algebraic system in which all the axioms of a ring except the associativity of the multiplication are postulated.

If in $(\mathscr{T}, +, \cdot)$ every subring generated by two elements is associative then $(\mathscr{T}, +, \cdot)$ is called an *alternative ring*.

It can be shown that $(\mathscr{T}, +, \cdot)$ is an alternative ring if and only if

$$(XX)Y = X(XY) \quad \text{and} \quad (YX)X = Y(XX)$$

for every element X and Y of $\mathscr{T}$.

If in $(\mathscr{T}, +, \cdot)$ every subring generated by one element is associative then $(\mathscr{T}, +, \cdot)$ is called a *power-associative* ring.

It can be shown that if the additive part $(\mathscr{T}, +)$ of $(\mathscr{T}, +, \cdot)$ has no non-zero element A such that $nA = 0$ for some positive natural number n then $(\mathscr{T}, +, \cdot)$ is a power-associative ring if and only if

$$(XX)X = X(XX) \quad \text{and} \quad ((XX)X)X = (XX)(XX)$$

for every element X of $\mathscr{T}$.

Clearly, every ring is both an alternative and power-associative ring.

If $(\mathscr{T}, +, \cdot)$ is such that

$$X^2 = 0 \quad \text{and} \quad (XY)Z + (YZ)X + (ZX)Y = 0$$

for every element X, Y and Z of $\mathscr{T}$ then $(\mathscr{T}, +, \cdot)$ is called a *Lie ring*.

Clearly, every zero ring is a Lie ring. Moreover, if $\mathscr{T}$ is a Lie ring then

$XY = -YX$ for every element X and Y of $\mathcal{T}$. Thus, in a Lie ring the multiplication is *anticommutative*.

If $(\mathcal{T}, +, \cdot)$ is such that

$$XY = YX \qquad \text{and} \qquad X^2(YX) = (X^2Y)X$$

for every element X and Y of $\mathcal{T}$ then $(\mathcal{T}, +, \cdot)$ is called a *Jordan ring*.

Clearly, every commutative ring is a Jordan ring.

In connection with nonassociative algebraic systems, let us mention the set of all *Cayley numbers* which is an alternative division ring under the ordinary addition and multiplication of Cayley numbers.

Exercises

1. Prove or disprove that the set of all elements A of a ring $\mathcal{R}$ such that $AX + Y + AXY = 0$ for every element X of $\mathcal{R}$ and some element Y of $\mathcal{R}$ is an ideal of $\mathcal{R}$.

2. Let A be an element of a ring $\mathcal{R}$. Prove that $A + B - AB = 0$ for some element B of $\mathcal{R}$ if and only if the set of all elements of the form $AX - X$ with $X \in \mathcal{R}$ is equal to $\mathcal{R}$.

3. Prove that if L is a left unity of a ring $\mathcal{R}$ and $A \in \mathcal{R}$ then $AL + L - A$ is also a left unity of $\mathcal{R}$. Prove an analogous assertion for a right unity of $\mathcal{R}$.

4. Prove that a ring with one and only one left unity is a ring with unity.

5. Let A and B be elements of a ring $\mathcal{R}$. Prove that the set $\mathcal{S}$ of all elements of the form AXB with $X \in \mathcal{R}$ is a subring of $\mathcal{R}$. Give some nontrivial conditions under which $\mathcal{S}$ will be an ideal of $\mathcal{R}$.

6. Prove that in a ring with unity an element has either one right inverse or infinitely many right inverses or no right inverse.

7. Let $\mathcal{A}$ and $\mathcal{B}$ be two ideals of a ring $\mathcal{R}$ and $\mathcal{A} + \mathcal{B}$ denote the set of all elements of $\mathcal{R}$ each of the form $X + Y$ with X an element of $\mathcal{A}$ and Y an element of $\mathcal{B}$. Also, let $\mathcal{A}\mathcal{B}$ denote the set of all elements of $\mathcal{R}$ each of which is a finite sum of summands each of the form XY with X an element of $\mathcal{A}$ and Y an element of $\mathcal{B}$.

Prove that $\mathcal{A} + \mathcal{B}$ as well as $\mathcal{A}\mathcal{B}$ is an ideal of $\mathcal{R}$. Moreover, prove that $\mathcal{A}\mathcal{B} \subset \mathcal{A} \cap \mathcal{B}$ and that $\mathcal{A}(\mathcal{B} + \mathcal{C}) = \mathcal{A}\mathcal{B} + \mathcal{A}\mathcal{C}$ where $\mathcal{C}$ is also an ideal of $\mathcal{R}$.

8. Let $\mathcal{A}$, $\mathcal{B}$ and $\mathcal{C}$ be ideals of a ring $\mathcal{R}$. Prove that if $\mathcal{A} \subset \mathcal{C}$ then $\mathcal{A} + (\mathcal{B} \cap \mathcal{C}) = (\mathcal{A} + \mathcal{B}) \cap \mathcal{C}$.

9. Let $\mathcal{R}$ be a commutative ring such that $\mathcal{R}\mathcal{R} = \mathcal{R}$. Prove that every maximal ideal of $\mathcal{R}$ is also a prime ideal of $\mathcal{R}$.

10. Let $\mathcal{M}$ be a maximal ideal of a ring $\mathcal{R}$ and $\mathcal{A}$ and $\mathcal{B}$ ideals of $\mathcal{R}$ neither of which is a subset of $\mathcal{M}$. Prove that $\mathcal{M} + \mathcal{A} = \mathcal{M} + \mathcal{B} = (\mathcal{M} + \mathcal{A})(\mathcal{M} + \mathcal{B}) = \mathcal{R}$.

11. Prove that an ideal $\mathcal{P}$ of a commutative ring $\mathcal{R}$ is a prime ideal if and only if $\mathcal{A}\mathcal{B} \subset \mathcal{P}$ implies $\mathcal{A} \subset \mathcal{P}$ or $\mathcal{B} \subset \mathcal{P}$ for every (left, right) ideal $\mathcal{A}$ and $\mathcal{B}$ of $\mathcal{R}$.

12. Let $\mathscr{P}$ be an ideal of a ring $\mathscr{R}$ such that $\mathscr{A}\mathscr{B} \subset \mathscr{P}$ implies $\mathscr{A} \subset \mathscr{P}$ or $\mathscr{B} \subset \mathscr{P}$ for every (left, right) ideal $\mathscr{A}$ and $\mathscr{B}$ of $\mathscr{R}$. Prove that $\mathscr{P}$ is not necessarily a prime ideal of $\mathscr{R}$ in the sense of the definition of a prime ideal given in the text.

13. Prove that in a commutative ring with unity every maximal ideal of the ring is a prime ideal. Prove also that in a finite commutative ring with unity an ideal different from the ring is a prime ideal if and only if it is a maximal ideal.

14. Let $\mathscr{R}$ be a commutative ring and $\mathscr{M}$ an ideal of $\mathscr{R}$. Prove that the quotient ring $\mathscr{R}/\mathscr{M}$ is a field if and only if $\mathscr{M}$ is a maximal ideal of $\mathscr{R}$ and $X^2 \in \mathscr{M}$ implies $X \in \mathscr{M}$ for every element X of $\mathscr{R}$.

15. Prove that the quotient ring $\mathscr{B}/\mathscr{M}$ is a Boolean ring if $\mathscr{M}$ is an ideal of a Boolean ring $\mathscr{B}$. Moreover, $\mathscr{B}/\mathscr{M}$ has two elements if and only if $\mathscr{M}$ is a maximal ideal of $\mathscr{B}$ and $\mathscr{B}$ has at least two elements.

16. Prove that the set $\mathscr{N}$ of all nilpotent elements of a commutative ring $\mathscr{R}$ is an ideal of $\mathscr{R}$ and that the quotient ring $\mathscr{R}/\mathscr{N}$ has no nonzero nilpotent elements.

17. Using Zorn's lemma prove that in a commutative ring with unity every ideal different from the ring is contained in a maximal ideal.

18. Using Zorn's lemma prove that the intersection of all the prime ideals of a commutative ring is equal to the set of all nilpotent elements of the ring.

19. Let $(\mathscr{R}, +, \cdot)$ be a ring. Prove that $(\mathscr{R}, +, *)$ is a Lie ring if

$$X * Y = XY - YX$$

for every element X and Y of $\mathscr{R}$.

20. Let $(\mathscr{R}, +, \cdot)$ be a ring. Prove that $(\mathscr{R}, +, *)$ is a Jordan ring if

$$X * Y = XY + YX$$

for every element X and Y of $\mathscr{R}$.

1.3. Direct Sum and Tensor Product of Rings

There are many ways to form a new ring from one or more given rings and, conversely, to decompose judiciously a given ring into some component rings. In the former case the new rings may acquire some properties which are not shared by the given rings. In the latter case the component rings may be of simpler structure facilitating the study of the given ring.

We have already mentioned the quotient ring $\mathscr{R}/\mathscr{I}$ of a ring $\mathscr{R}$ with respect to an ideal $\mathscr{I}$ of $\mathscr{R}$. As mentioned, for a particular choice of $\mathscr{I}$ the quotient ring $\mathscr{R}/\mathscr{I}$ may have some properties which are not enjoyed by $\mathscr{R}$.

Some of the standard procedures involving the above ideas consist of forming the direct sum of some given rings, decomposing a ring into a direct sum of some of its ideals, forming the tensor product of some given rings and decomposing a ring into a tensor product of some of its subrings.

Consider the rings $\mathscr{R}_1, \mathscr{R}_2, \ldots, \mathscr{R}_n$ and let $\mathscr{S}$ be the set of all ordered n-tuples $(A_1, A_2, \ldots, A_n)$ with $A_i \in \mathscr{R}_i$. Then the *direct sum*, or more appropriately, the *external direct sum*

$$\mathscr{R}_1 \oplus \mathscr{R}_2 \oplus \cdots \oplus \mathscr{R}_n = \bigoplus_{i=1}^{n} \mathscr{R}_i \qquad (11)$$

of the rings $\mathscr{R}_1, \mathscr{R}_2, \ldots, \mathscr{R}_n$ is defined as the set $\mathscr{S}$ where addition and multiplication among the elements of $\mathscr{S}$ are performed coordinatewise (or componentwise). In (11) each $\mathscr{R}_i$ is called a *component* or a *summand* or, more precisely, a *direct summand* of $\bigoplus_{i=1}^{n} \mathscr{R}_i$.

It is readily seen that (11) with addition and multplication performed coordinatewise is a ring and that $(0_1, 0_2, \ldots, 0_n)$ is the zero of (11) where 0_i is the zero of $\mathscr{R}_i$. Moreover, (11) has a unit element I if and only if each $\mathscr{R}_i$ has a unit element I_i. Clearly, $I = (I_1, I_2, \ldots, I_n)$.

One can easily verify that a direct sum of rings is commutative if and only if each summand is commutative. Moreover, a direct sum has a divisor of zero if two of its summands each has more than one element. Thus, in particular, a direct sum of two fields is not even an integral domain.

The concept of the direct sum of finitely many rings, as described above, can be generalized to the case of infinitely many rings in at least two obvious ways. Let $(\mathscr{R}_i)_{i \in \alpha}$ be a family of rings. Consider the set $\mathscr{S}$ of all families $(A_i)_{i \in \alpha}$ with $A_i \in \mathscr{R}_i$ and the set $\mathscr{F}$ of all families $(A_i)_{i \in \alpha}$ with $A_i \in \mathscr{R}_i$ and $A_i \neq 0_i$ for only finitely many i's. Clearly, $\mathscr{S}$ as well as $\mathscr{F}$ is a ring if addition and multiplication among their respective elements are performed coordinatewise. In this case $\mathscr{S}$ is called the *direct sum* and $\mathscr{F}$ the *discrete direct sum* of the family of rings $(\mathscr{R}_i)_{i \in \alpha}$.

Consider (11) and let $\mathscr{R}_i'$ for $i = 1, 2, \ldots, n$ denote the set of all n-tuples $(0_1, 0_2, \ldots, A_i, \ldots, 0_n)$ with $A_i \in \mathscr{R}_i$. Then $\mathscr{R}_i'$ is a subring (in fact an ideal) of (11) and is isomorphic to the ring $\mathscr{R}_i$. Moreover,

$$\left(\bigoplus_{i=1}^{n} \mathscr{R}_i \right) / \mathscr{R}_i' \cong \bigoplus_{j \neq i} \mathscr{R}_j. \qquad (12)$$

Let $\mathscr{R}_i'$ be defined above. Then in connection with external direct sum (11) we make the following two observations:

Every element of $\bigoplus_{i=1}^{n} \mathscr{R}_i$ *is uniquely represented as a sum of* $\qquad (13)$
elements each belonging to a distinct $\mathscr{R}_i'$ *for* $i = 1, 2, \ldots, n$

For every $A_i' \in \mathcal{R}_i'$ and $A_j' \in \mathcal{R}_j'$, with $i, j = 1, 2, \ldots, n$ $\qquad$ (14)

$$A_i' A_j' = (0_1, 0_2, \ldots, p_n), \qquad if \qquad i \neq j$$

Motivated by (11), (13) and (14) we say that a ring $\mathcal{R}$ is the *direct sum* (or, more appropriately, *the internal direct sum*) of its subrings $\mathcal{R}_1$, $\mathcal{R}_2, \ldots, \mathcal{R}_n$ and we denote this by

$$\mathcal{R} = \mathcal{R}_1 \oplus \mathcal{R}_2 \oplus \cdots \oplus \mathcal{R}_n = \overset{n}{\underset{i=1}{\oplus}} \mathcal{R}_i \qquad (15)$$

if the following two conditions are satisfied:

Every element of $\mathcal{R}$ is uniquely represented as a sum of ele- $\qquad$ (16)
ments each belonging to a distinct $\mathcal{R}_i$ for $i = 1, 2, \ldots, n$.

For every $A_i \in \mathcal{R}_i$ and $A_j \in \mathcal{R}_j$, with $i, j = 1, 2, \ldots, n$ $\qquad$ (17)

$$A_i A_j = 0, \qquad if \qquad i \neq j$$

where 0 is naturally the zero of the ring $\mathcal{R}$.

In view of (17) it is customary to say that the distinct summands of direct sum (15) are *pairwise orthogonal*.

Let us observe that although the term *direct sum* and the symbol $\oplus$ are used in connection with either external or internal direct sum, the context will prevent any ambiguity.

Next, we introduce some definitions and notations which are justified since addition and multiplication in a ring are associative binary operations.

Let $\mathcal{S}_1, \mathcal{S}_2, \ldots, \mathcal{S}_n$ be subsets of a ring $(\mathcal{R}, +, \cdot)$. We define their *sum*, denoted by

$$\mathcal{S}_1 + \mathcal{S}_2 + \cdots + \mathcal{S}_n = \sum_{i=1}^{n} \mathcal{S}_i \qquad (18)$$

as the set of all sums $A_1 + A_2 + \cdots + A_n = \sum_{i=1}^{n} A_i$ with $A_i \in \mathcal{S}_i$ for $i = 1, 2, \ldots, n$. Moreover, we define their *product* denoted by

$$\mathcal{S}_1 \mathcal{S}_2 \cdots \mathcal{S}_n = \prod_{i=1}^{n} \mathcal{S}_i \qquad (19)$$

as the set of all finite sums of products $A_1 A_2 \cdots A_n = \prod_{i=1}^{n} A_i$ with $A_i \in \mathcal{S}_i$ for $i = 1, 2, \ldots, n$.

In particular, for every subset $\mathcal{S}$ of $\mathcal{R}$ and every element A of $\mathcal{R}$ we define

$$\mathcal{S}\{A\} = \mathcal{S}A \qquad and \qquad \{A\}\mathcal{S} = A\mathcal{S} \qquad (20)$$

and we observe that if $\mathscr{S}$ is a subgroup of $\mathscr{R}$ then $\mathscr{S}A$ is the set of all elements of $\mathscr{R}$ of the form SA with $S \in \mathscr{S}$. Similarly, if $\mathscr{S}$ is a subgroup of $\mathscr{R}$ then $A\mathscr{S}$ is the set of all elements of $\mathscr{R}$ of the form AS with $S \in \mathscr{S}$.

Clearly, for every subset $\mathscr{S}$ of $\mathscr{R}$ we have:

$$\mathscr{S} \sum_{i=1}^{n} \mathscr{S}_i = \sum_{i=1}^{n} \mathscr{S}\mathscr{S}_i \qquad \text{and} \qquad \left(\sum_{i=1}^{n} \mathscr{S}_i\right)\mathscr{S} = \sum_{i=1}^{n} \mathscr{S}_i\mathscr{S}, \tag{21}$$

$$\mathscr{S} \bigcap_{i=1}^{n} \mathscr{S}_i \subset \bigcap_{i=1}^{n} \mathscr{S}\mathscr{S}_i \qquad \text{and} \qquad \left(\bigcap_{i=1}^{n} \mathscr{S}_i\right)\mathscr{S} \subset \bigcap_{i=1}^{n} \mathscr{S}_i\mathscr{S}, \tag{22}$$

$$\mathscr{S} + \bigcap_{i=1}^{n} \mathscr{S}_i \subset \bigcap_{i=1}^{n} (\mathscr{S} + \mathscr{S}_i) \qquad \text{and} \qquad \left(\bigcap_{i=1}^{n} \mathscr{S}_i\right) + \mathscr{S} \subset \bigcap_{i=1}^{n} (\mathscr{S}_i + \mathscr{S}). \tag{23}$$

Now, if each $\mathscr{S}_i$ in the above is an ideal $\mathscr{I}_i$ of the ring $(\mathscr{R}, +, \cdot)$ then $\sum_{i=1}^{n} \mathscr{I}_i$ and $\prod_{i=1}^{n} \mathscr{I}_i$ as well as $\bigcap_{i=1}^{n} \mathscr{I}_i$ is an ideal of $\mathscr{R}$. Moreover,

$$\prod_{i=1}^{n} \mathscr{I}_i \subset \bigcap_{i=1}^{n} \mathscr{I}_i \subset \sum_{i=1}^{n} \mathscr{I}_i, \tag{24}$$

LEMMA 1. *Let $\mathscr{R}_1, \mathscr{R}_2, \ldots, \mathscr{R}_n$ be subrings of a ring $\mathscr{R}$. Then (25) through (28) are equivalent statements.*

$$\mathscr{R} = \bigoplus_{i=1}^{n} \mathscr{R}_i. \tag{25}$$

Each $\mathscr{R}_i$ is an ideal of $\mathscr{R}$ and each element of $\mathscr{R}$ is uniquely (26) *represented as a sum of elements each belonging to a distinct $\mathscr{R}_i$, for $i = 1, 2, \ldots, n$.*

$\mathscr{R} = \sum_{i=1}^{n} \mathscr{R}_i$, *each $\mathscr{R}_i$ is an ideal of $\mathscr{R}$ and $\sum_{i=1}^{n} A_i = 0$ implies* (27)

$A_i = 0$, *for $A_i \in \mathscr{R}_i, i = 1, 2, \ldots, n$.*

$\mathscr{R} = \sum_{i=1}^{n} \mathscr{R}_i$, *each $\mathscr{R}_i$ is an ideal of $\mathscr{R}$ and $\mathscr{R}_i \cap \sum_{j \neq i} \mathscr{R}_j = \{0\}$, for* (28) *$i = 1, 2, \ldots, n$.*

Proof. Let (25) hold and let $A_i \in \mathscr{R}_i$ and $A \in \mathscr{R}$. By (16) we have $A = A_1 + \cdots + A_i + \cdots + A_n$. By (17) we have $AA_i = A_iA_i$ and $A_iA = A_iA_i$. However, $\mathscr{R}_i$ is a subring of $\mathscr{R}$ and thus, AA_i as well as A_iA is an element of $\mathscr{R}_i$. Consequently, $\mathscr{R}_i$ is an ideal of $\mathscr{R}$ which in view of (16) establishes (26).

Next, let (26) hold and let $A_1 + A_2 + \cdots + A_n = 0$ with A_i an element of the ideal $\mathscr{R}_i$. Since $0 + 0 + \cdots + 0 = 0$ we see that $A_i = 0$ for $i = 1, 2, \ldots, n$ which establishes (27).

Next, let (27) hold and let $A_i \in \mathcal{R}_i$ and $A_i \in \sum_{j \neq i} \mathcal{R}_j$. Thus, $A_i - \sum_{j \neq i} A_j = 0$ and since $\mathcal{R}_i$ is an ideal of $\mathcal{R}$ we see that $A_i = 0$ which establishes (28).

Finally, let (28) hold and let $A \in \mathcal{R}$. Also, $A = A_1 + A_2 + \cdots + A_n = A_1' + A_2' + \cdots + A_n'$. But then $A_i - A_i' = \sum_{j \neq i} (A_j - A_j')$ which implies $A_i = A_i'$ which in turn implies (16). On the other hand, since $\mathcal{R}_i$ and $\mathcal{R}_j$ are ideals of $\mathcal{R}$ we have $A_i A_j \in \mathcal{R}_i$ and $A_j A_i \in \mathcal{R}_j$. However, if $i \neq j$ then in view of (24) we see that $A_i A_j = 0$ which implies (17). Hence (28) implies (25).

Thus Lemma 1 is proved.

LEMMA 2.   *Let $\mathcal{R}$ be a ring and $\mathcal{I}$ an ideal of $\mathcal{R}$ with a unit element E. Then $\mathcal{I}$ is a direct summand of $\mathcal{R}$, i.e.,*

$$\mathcal{R} = \mathcal{I} \oplus \mathcal{K} \tag{29}$$

for some ideal $\mathcal{K}$ of $\mathcal{R}$. Moreover, $\mathcal{K}$ is unique.

Proof.   Clearly, the set $\mathcal{K}$ of all elements X of $\mathcal{R}$ such that $XY = YX = 0$ for every element Y of $\mathcal{I}$ is an ideal of $\mathcal{R}$. Moreover, $A - AE$ is an element of $\mathcal{K}$ and $A = AE + (A - AE)$ for every element A of $\mathcal{R}$. Thus, $\mathcal{R} = \mathcal{I} + \mathcal{K}$. Now, if $A - AE = H$ for some element H of $\mathcal{I}$ then $0 = (A - AE)E = HE = H$. Hence, $H = 0$ which in view of (28) establishes (29).

The uniqueness of $\mathcal{K}$ follows from its definition and the fact that $\mathcal{I}$ has a unit element.

As shown by the lemmas below, rings with unity have some significant properties which are not shared by rings without unity.

In particular, Lemma 3 shows that the existence of the unit element in a ring together with condition (16) implies condition (15).

LEMMA 3.   *Let $\mathcal{R}$ be a ring with unit element I and let $\mathcal{R}_1, \mathcal{R}_2, \ldots, \mathcal{R}_n$ be subrings of $\mathcal{R}$ such that every element of $\mathcal{R}$ is uniquely represented as a sum of elements each belonging to a distinct $\mathcal{R}_i$ for $i = 1, 2, \ldots, n$. Then*

$$\mathcal{R} = \mathcal{R}_1 \oplus \mathcal{R}_2 \oplus \cdots \oplus \mathcal{R}_n.$$

Proof.   Let

$$I = I_1 + I_2 + \cdots + I_n \quad \text{with} \quad I_i \in \mathcal{R}_i.$$

Since $\mathcal{R}_i$ is a subring of $\mathcal{R}$, in view of the unique representation of every element of $\mathcal{R}$ as a sum of the elements of $\mathcal{R}_i$ we have

$$I_i I_j = 0 \quad \text{for} \quad i \neq j \quad \text{and} \quad I_i I_i = I_i$$

where 0 is the zero of $\mathcal{R}$. Clearly, I_i is the unity of $\mathcal{R}_i$. Consequently, for every $A_i \in \mathcal{R}_i$ and $A_j \in \mathcal{R}_j$ we have

$$A_i A_j = 0 \quad \text{if} \quad i \neq j$$

which in view of (17) implies the conclusion of the lemma.

LEMMA 4. *Let $\mathscr{R}$ be a ring with unit element I and*

$$\mathscr{R} = \mathscr{R}_1 \oplus \mathscr{R}_2 \oplus \cdots \oplus \mathscr{R}_n \qquad (30)$$

where $\mathscr{R}_i$ is an ideal of $\mathscr{R}$ for $i = 1, 2, \ldots, n$.
Then a subring $\mathscr{S}$ of $\mathscr{R}$ is an (left, right) ideal of $\mathscr{R}$ if and only if

$$\mathscr{S} = \mathscr{S}_1 \oplus \mathscr{S}_2 \oplus \cdots \oplus \mathscr{S}_n \qquad (31)$$

where $\mathscr{S}_i$ is an (left, right) ideal of $\mathscr{R}_i$ and $\mathscr{S}_i = \mathscr{S} \cap \mathscr{R}_i$.

Proof. Since $\mathscr{R}$ has a unit element $\mathscr{R}\mathscr{S} = \mathscr{S}$ and $\mathscr{S}\mathscr{R} = \mathscr{S}$ respectively for a left and right ideal $\mathscr{S}$ of $\mathscr{R}$. But then in view of (30) and (21) one can easily prove that,

$$\mathscr{S} = \mathscr{R}\mathscr{S} = \overset{n}{\underset{i=1}{\oplus}} \mathscr{R}_i\mathscr{S} \qquad \text{and} \qquad \mathscr{S} = \mathscr{S}\mathscr{R} = \overset{n}{\underset{i=1}{\oplus}} \mathscr{S}\mathscr{R}_i$$

respectively for a left and right ideal $\mathscr{S}$ of $\mathscr{R}$. Denoting in the above $\mathscr{R}_i\mathscr{S} = \mathscr{S} \cap \mathscr{R}_i$ by $\mathscr{S}_i$ and $\mathscr{S}\mathscr{R}_i = \mathscr{S} \cap \mathscr{R}_i$ by $\mathscr{S}_i$ respectively for a left and right ideal $\mathscr{S}$ of $\mathscr{R}$ we derive (31).

Next, let (31) hold. In view of (30) we have $I = I_1 + I_2 + \cdots + I_n$ where I_i is the unity of the ideal $\mathscr{R}_i$ of $\mathscr{R}$. But then $\mathscr{S} \cap \mathscr{R}_i = \mathscr{R}\mathscr{S}_i = \mathscr{R}_i\mathscr{S}_i = \mathscr{S}_i$ and $\mathscr{S} \cap \mathscr{R}_i = \mathscr{S}_i\mathscr{R} = \mathscr{S}\mathscr{R}_i = \mathscr{S}_i$ respectively for a left and right ideal $\mathscr{S}_i$ of $\mathscr{R}_i$. Consequently, in view of (30)

$$\mathscr{R}\mathscr{S} = \overset{n}{\underset{i=1}{\oplus}} \mathscr{R}_i\mathscr{S}_i = \overset{n}{\underset{i=1}{\oplus}} \mathscr{S}_i = \mathscr{S}$$

and

$$\mathscr{S}\mathscr{R} = \overset{n}{\underset{i=1}{\oplus}} \mathscr{S}_i\mathscr{R}_i = \overset{n}{\underset{i=1}{\oplus}} \mathscr{S}_i = \mathscr{S}$$

respectively for a left and right ideal $\mathscr{S}_i$ of $\mathscr{R}_i$ which implies that $\mathscr{S}$ is respectively a left and right ideal of $\mathscr{R}$.

LEMMA 5. *Let $\mathscr{R}$ be a ring and $\mathscr{C}$ the center of $\mathscr{R}$. Then*

$$\mathscr{R} = \mathscr{R}_1 \oplus \mathscr{R}_2 \oplus \cdots \oplus \mathscr{R}_n \qquad (32)$$

implies

$$\mathscr{C} = \mathscr{C}_1 \oplus \mathscr{C}_2 \oplus \cdots \oplus \mathscr{C}_n \qquad (33)$$

where $\mathscr{C}_i$ is the center of $\mathscr{R}_i$ and $\mathscr{C}_i = \mathscr{C}_i = \mathscr{C} \cap \mathscr{R}_i$.
Moreover, if $\mathscr{R}$ has a unit element then (33) implies (32).

Proof. Let (32) hold. To prove (33) it is enough to choose $\mathscr{C}_i = \mathscr{C} \cap \mathscr{R}_i$. Next, let $\mathscr{R}$ have a unit and let (33) hold. To prove (32) it is enough to choose $\mathscr{R}_i = \mathscr{R}\mathscr{C}_i = \mathscr{C} \cap \mathscr{R}_i$.

If each $\mathscr{R}_i$ for $i = 1, 2, \ldots, n$ is a subring of a ring the product $\Pi \mathscr{R}_i$, as given by (19), need not, in general, be a subring of $\mathscr{R}$. However, this is not the case in connection with the notion of the tensor product of subrings introduced below.

Let $\mathscr{R}_i, \mathscr{R}_2, \ldots, \mathscr{R}_n$ be subrings of a ring such that for $i, j = 1, 2, \ldots, n$.

$$A_i A_j = A_j A_i \tag{34}$$

where $A_i \in \mathscr{R}_i$ and $A_j \in \mathscr{R}_j$. Let $\mathscr{P}$ be the set of all finite sums of the products $A_1 A_2 \ldots A_n = \prod_{i=1}^{n} A_i$ with $A_i \in \mathscr{R}_i$. Then the *tensor product* (or, more appropriately, the *internal tensor product*)

$$\mathscr{R}_1 \otimes \mathscr{R}_2 \otimes \cdots \otimes \mathscr{R}_n = \overset{n}{\underset{i=1}{\otimes}} \mathscr{R}_i \tag{35}$$

of the subrings $\mathscr{R}_1, \mathscr{R}_2, \ldots, \mathscr{R}_n$ of $\mathscr{R}$ is defined as the set $\mathscr{P}$ where addition and multiplication among the elements of $\mathscr{P}$ are performed according to the addition and multiplication in $\mathscr{R}$. In (35) each $\mathscr{R}_i$ is called a *tensor factor* of (35).

In view of (34) it is readily seen that (35) is a ring or, more precisely, a subring of $\mathscr{R}$. Clearly, $\mathscr{P} = \prod_{i=1}^{n} \mathscr{R}_i$. Thus, the tensor product $\overset{n}{\underset{i=1}{\otimes}} \mathscr{R}_i$ of the subrings $\mathscr{R}_i$ of the ring $\mathscr{R}$ is a special case of the product $\prod_{i=1}^{n} \mathscr{R}_i$, as given by (19), where condition (34) is satisfied.

As expected, 0 of $\mathscr{R}$ is the zero of (35) and if each $\mathscr{R}_i$ has a unity I_i then $I_1 I_2 \ldots I_n$ is the unity of (35).

We observe that in the above we assumed, to begin with, that each $\mathscr{R}_i$ with $i = 1, 2, \ldots, n$ is a subring of the ring $\mathscr{R}$.

Let us mention only that it is possible to define the tensor product of rings $\mathscr{R}_i$ where none of $\mathscr{R}_i$ is a subring of a given ring $\mathscr{R}$. Also, the concept of the tensor product of rings can be generalized in many ways to the case of infinitely many factors. However, we shall not have the occasion to deal with these items.

Motivated by (35) and, as expected, we say that a ring $\mathscr{R}$ is the *tensor product* of its subrings $\mathscr{R}_1, \mathscr{R}_2, \ldots, \mathscr{R}_n$ if

$$\mathscr{R} = \mathscr{R}_1 \otimes \mathscr{R}_2 \otimes \cdots \otimes \mathscr{R}_n.$$

Later on we shall see that the concept of the tensor product of rings is used to define the concept of the tensor product of algebras.

Exercises

1. Prove (12), (21), (22), (23) and (24).

2. Prove that a finite zero ring is a direct sum of some of its ideals each having a prime power number of elements.

3. Let $\mathscr{R}$ be a ring whose characteristic is a product of two relatively prime natural numbers n_1 and n_2 each greater than 1. Prove or disprove that $\mathscr{R}$ is a direct sum of two of its ideals whose characteristics are respectively n_1 and n_2.

4. Let $\mathscr{R}$ be a commutative ring. Prove that $\mathscr{R}$ is isomorphic to a subring of a direct sum of fields if and only if $\mathscr{R}$ has no nonzero nilpotent elements.

5. Let $\mathscr{R}$ be a commutative ring with more than one element. Prove that $\mathscr{R}$ is a direct sum of finitely many of its subfields if and only if $\mathscr{R}$ has finitely many ideals and has no nonzero nilpotent element.

6. A subring $\mathscr{S}$ of a direct sum of a family $(\mathscr{R}_i)_{i\epsilon\alpha}$ of rings $\mathscr{R}_i$ is called a *subdirect sum* of the family $(\mathscr{R}_i)_{i\epsilon\alpha}$ if every element of every $\mathscr{R}_i$ is used as a coordinate of some element of $\mathscr{S}$. Prove that a commutative ring has no nonzero nilpotent element if and only if it is isomorphic to a subdirect sum of integral domains.

7. Let $(\mathscr{I}_i)_{i\epsilon\alpha}$ be a family of ideals of a ring $\mathscr{R}$ such that $\bigcap_{i\epsilon\alpha} \mathscr{I}_i = \{0\}$. Prove that $\mathscr{R}$ is isomorphic to a subdirect sum (*see* Problem 6) of the quotient rings $\mathscr{R}/\mathscr{I}_i$ with $i\epsilon\alpha$.

8. Let $\mathscr{M}_1, \mathscr{M}_2, \ldots, \mathscr{M}_n$ be maximal ideals of a ring $\mathscr{R}$ such that their intersection is $\{0\}$ and such that the intersection of any $n-1$ of them is not $\{0\}$. Prove that $\mathscr{R}$ is isomorphic to the direct sum of the quotient rings $\mathscr{R}/\mathscr{M}_i$ with $i = 1, 2, \ldots, n$.

9. Let $\mathscr{R}$ be a commutative ring with more than one element. Let also $\mathscr{A}$ be the set of all elements A of $\mathscr{R}$ such that $AX + Y + AXY = 0$ for every element X of $\mathscr{R}$ and some element Y of $\mathscr{R}$. Prove or disprove that $\mathscr{R}$ is isomorphic to a subdirect sum (*see* Problem 6) of fields if and only if $\mathscr{A} = \{0\}$.

10. Prove that a finite Boolean ring (*i.e.*, a ring every element of which is idempotent) with more than one element is a direct sum of finitely many of its two-element subfields.

11. Prove that a ring $\mathscr{R}$ is isomorphic to a subdirect sum (*see* Problem 6) of two-element fields if and only if $\mathscr{R}$ is a Boolean ring with more than one element.

12. Let $\mathscr{R}$ be a commutative ring with more than one element such that $XYX = X$ for every element X of $\mathscr{R}$ and some element Y of $\mathscr{R}$. Prove or disprove that $\mathscr{R}$ is isomorphic to a subdirect sum (*see* Problem 6) of fields.

13. Let p be a prime number. A ring $\mathscr{R}$ is called a p-ring if $X^p = X$ and $pX = 0$ for every element X of $\mathscr{R}$. Prove that a finite p-ring with more than one element is a direct sum of finitely many of its p-element subfields.

14. Let p be a prime number. Prove that a ring $\mathscr{R}$ is isomorphic to a subdirect sum (*see* Problem 6) of p-element fields if and only if $\mathscr{R}$ is a p-ring (*see* Problem 11) with more than one element.

15. Let p be a prime number and k a natural number greater than or equal to 1. Let $\mathscr{R}$ be a commutative ring such that $X^{p^k}Y = XY^{p^k}$ and $pX = 0$ for every element

X and Y of $\mathscr{R}$. Prove that $\mathscr{R}$ is a direct sum of two of its ideals $\mathscr{I}_1$ and $\mathscr{I}_2$ such that $X_{p^k} = X$ for every element X of $\mathscr{I}_1$ and that every element of $\mathscr{I}_2$ is nilpotent.

16. Let p be a prime number and let $\mathscr{R}$ be a ring such that $X^p Y = X Y^p$ and $pX = 0$ for every element X and Y of $\mathscr{R}$. Prove or disprove that $\mathscr{R}$ is a direct sum of two of its ideals $\mathscr{I}_1$ and $\mathscr{I}_2$ such that $X^p = X$ for every element X of $\mathscr{I}_1$ and that every element of $\mathscr{I}_2$ is nilpotent.

1.4. Polynomial Rings

Besides forming direct sums and tensor products of rings there are other standard methods of forming new rings based on the structure of a given ring. One of these methods consists of forming polynomial rings over a given ring (*i.e.*, with coefficients in a given ring).

So far as polynomial rings are concerned we shall consider mainly polynomials in one indeterminate (or variable) with coefficients in a ring $\mathscr{R}$. In this connection we shall denote the elements of $\mathscr{R}$ by small letters. In particular, we denote the unity of $\mathscr{R}$ by 1, if it exists.

Let $(\mathscr{R}, +, \cdot)$ be a ring and let $\mathscr{P}$ be the set of all sequences

$$(a_i) = (a_0, a_1, a_2, \ldots) \qquad (a_i \in \mathscr{R})$$

where *all but a finite number of a_i are* 0.

As expected, we define equality between two elements (a_i) and (b_i) of $\mathscr{P}$ by

$$(a_i) = (b_i) \quad \text{if} \quad a_i = b_i \quad (i = 0, 1, 2, \ldots) \tag{36}$$

In $\mathscr{P}$ we define addition as

$$(a_i) + (b_i) = (a_i + b_i) \qquad (i = 0, 1, 2, \ldots) \tag{37}$$

and we define multiplication as

$$(a_i)(b_i) = (c_i) \tag{38}$$

where

$$c_i = \sum_{m+n=i} a_m b_n \qquad (i = 0, 1, 2, \ldots.)$$

In view of (36), (37) and (38) it can be readily verified that $(\mathscr{R}, +, \cdot)$ is a ring. This ring is called a *polynomial ring over the ring $\mathscr{R}$* and an element of $\mathscr{P}$ is called a *polynomial over the ring $\mathscr{R}$*.

Clearly, the zero of $\mathscr{P}$ is $(0, 0, 0, \ldots)$. Moreover, $\mathscr{P}$ has a unity if and only if $\mathscr{R}$ has a unity 1 in which case the unity of $\mathscr{P}$ is $(1, 0, 0, \ldots)$.

Although the theory of polynomial rings can be developed based on the above definition and notation of a polynomial, we shall not pursue this course. Instead, first, we shall represent the polynomial $(a_0, a_1, a_2, \ldots)$ by a_0 if $a_n = 0$ for $n > 0$; otherwise, by

$$a_n X^n + a_{n-1} X^{n-1} + \cdots + a_1 X^1 + a_0 X^0$$

where a_n is the last nonzero coordinate in $(a_0, a_1, a_2, \ldots)$ with $n > 0$. Next, we shall write a_0 instead of $a_0 X^0$ and, as usual, we shall write X instead of X^1.

Thus, from now on by a polynomial over a ring $\mathscr{R}$ we shall mean a finite expression of the form a_0 if $a_n = 0$ for $n > 0$; otherwise,

$$a_n X^n + \cdots + a_1 X + a_0 \quad \text{with} \quad a_i \in \mathscr{R} \quad \text{and} \quad a_n \neq 0 \text{ with } n > 0 \quad (39)$$

where X is any symbol which is not an element of $\mathscr{R}$ and which is called an *indeterminate*.

As usual, we shall refer to (39) as a *polynomial in one indeterminate X over the ring $\mathscr{R}$*. In this connection we observe that every element of $\mathscr{R}$ is a polynomial which will be called a *constant polynomial*. In particular the zero 0 of $\mathscr{R}$ is called the *zero polynomial* and if $\mathscr{R}$ has a unity 1 then 1 is called the *unity polynomial*.

In a nonconstant polynomial the exponent of the highest power of X is called the *degree* of the polynomial and the corresponding coefficient is called the *leading coefficient* of the polynomial. We assign no degree to the zero polynomial and we assign the zero degree to every nonzero constant polynomial. Moreover, in a constant polynomial a_0 we call a_0 the *leading coefficient* of the polynomial a_0. If the leading coefficient of a polynomial is the unity of the corresponding ring then the polynomial is called a *monic polynomial*.

Let us denote the set of all expressions of the form (39) by $\mathscr{R}[X]$ and let us define two elements of $\mathscr{R}[X]$ equal if they contain precisely the same terms, except for terms with zero coefficients which may be excluded or included at will.

Without further comment we observe that $\mathscr{R}[X]$ is a ring with addition and multiplication performed among elements of $\mathscr{R}[X]$ according to the high school method except for preserving the obvious order in forming the coefficients of the product of two polynomials and for writing the coefficient a_i of X^i in any polynomial to the left of X^i. As expected, we identify $1X$ with X if 1 is the unity of $\mathscr{R}$. Clearly, the zero of $\mathscr{R}[X]$ is the zero of $\mathscr{R}$ and the unity of $\mathscr{R}[X]$ is the unity of $\mathscr{R}$ if it exists. We observe also that $\mathscr{R}$ is a subring of $\mathscr{R}[X]$ and if $\mathscr{R}$ has a unity then X is an element of $\mathscr{R}[X]$.

We call $\mathscr{R}[X]$ *the ring of all polynomials in one indeterminate X over the ring $\mathscr{R}$*. It can be readily verified that, in general, $\mathscr{R}[X]$ is isomorphic to the ring $\mathscr{P}$ mentioned above.

Some of the structural relationships between the rings $\mathscr{R}$ and $\mathscr{R}[X]$ are as follows:

(i) The zero of $\mathscr{R}[X]$ is the zero of $\mathscr{R}$.

(ii) 1 is the unity of $\mathscr{R}[X]$ if and only if 1 is the unity of $\mathscr{R}$.

(iii) $\mathscr{R}[X]$ is commutative if and only if $\mathscr{R}$ is commutative.

(iv) $\mathscr{R}[X]$ is an integral domain if and only if $\mathscr{R}$ is an integral domain.

(v) $\mathscr{R}[X]$ is a Gaussian ring (*see* page 44, vol. 1. *Modern Algebra* by B. L. Van der Waerden) if and only if $\mathscr{R}$ is a Gaussian ring.

Consider the polynomial

$$P(X) = a_n X^n + \cdots + a_1 X + a_0$$

over a ring $\mathscr{R}$. Let

$$P_R(r) = a_n r^n + \cdots + a_1 r + a_0 \tag{40}$$

and

$$P_L(r) = r^n a_n + \cdots + r a_1 + a_0 \tag{41}$$

where $r \in \mathscr{R}$. We call $P_R(r)$ the *right-value* of $P(X)$ for $X = r$ and $P_L(r)$ the *left-value* of $P(X)$ for $X = r$. Clearly, $P_R(r)$ as well as $P_L(r)$ is an element of $\mathscr{R}$.

If $\mathscr{R}$ is a commutative ring then $P_R(r) = P_L(r)$ for every $P(X) \in \mathscr{R}[X]$ and every $r \in \mathscr{R}$. In this case we write

$$P_R(r) = P_L(r) = P(r)$$

and we call $P(r)$ the *value* of $P(X)$ for $X = r$. We observe that if $\mathscr{R}$ is a commutative ring then the mapping $P(X) \to P(r)$ is a homomorphism of the ring $\mathscr{R}[X]$ into the ring $\mathscr{R}$.

If $P(X)$ is a nonzero polynomial over a commutative ring $\mathscr{R}$ and if $P(r) = 0$ then we call r a *root* of the polynomial $P(X)$. In this connection let us mention that an element r' of a commutative ring $\mathscr{R}'$ is called *algebraic* over a subring $\mathscr{R}$ of $\mathscr{R}'$ if r' is a root of a nonzero element of $\mathscr{R}[X]$. Otherwise, r' is called *transcendental* over $\mathscr{R}$.

The concept of divisibility can be introduced in any ring $\mathscr{R}$. Thus, an element a of $\mathscr{R}$ is called *right divisible* by an element b of $\mathscr{R}$ if $a = cb$ for some element c of $\mathscr{R}$. Likewise, a is called *left divisible* by b if $a = be$ for some element e of $\mathscr{R}$. In the former case b is called a *right divisor* of a and in the latter case b is called a *left divisor* of a.

In the case of the ring $\mathscr{R}[X]$ the concept of divisibility can be refined to the so-called *Euclidean algorithm* for division which is described below.

EUCLIDEAN ALGORITHM. *Let $\mathscr{R}$ be a ring with unity and $P(X)$ and $G(X)$ elements of $\mathscr{R}[X]$ respectively of degree m and n and such that* *the leading coefficient of $G(X)$ has an inverse in $\mathscr{R}$. Then there exist unique elements $Q(X), F(X), R(X)$ and $S(X)$ of $\mathscr{R}[X]$ such that*

$$P(X) = Q(X)G(X) + R(X) \tag{42}$$

and

$$P(X) = G(X)F(X) + S(X) \tag{43}$$

where $R(X)$ and $S(X)$ are zero or of degree less than n and $Q(X)$ and $F(X)$ are zero or of degree $m - n$ if $m \geqslant n$.

Clearly, if $R(X) = 0$ in (42) then $G(X)$ is a right divisor of $P(X)$. Likewise, if $S(X) = 0$ then $G(X)$ is a left divisor of $P(X)$.

LEMMA 6. *Let $\mathscr{R}$ be a ring with unity and $a \in \mathscr{R}$ and $P(X) \in \mathscr{R}[X]$. Then*

$$P(X) = Q(X)(X - a) + P_R(a) \tag{44}$$

and

$$P(X) = (X - a)F(X) + P_L(a) \tag{45}$$

where $P_R(a)$ and $P_L(a)$ are as given by (40) and (41) respectively.

Proof.   In view of (42) we have

$$P(X) = Q(X)(X - a) + r$$

where r is an element of $\mathscr{R}$. But then if we choose for $Q(X)$ a polynomial $q_n X^n + \cdots + q_1 X + q_0$ a straightforward computation will show that $r = P_R(a)$, as desired. Likewise, (45) follows from (43).

As an immediate consequence of Lemma 6 we have:

LEMMA 7.   *Let $\mathscr{R}$ be a ring with unity and $a \in \mathscr{R}$ and $P(X) \in \mathscr{R}[X]$. Then $(X - a)$ is a right (or left) divisor of $P(X)$ if and only if $P_R(a) = 0$ (or $P_L(a) = 0$).*

Based on Lemma 7 one can prove easily that *if $\mathscr{R}$ is an integral domain then an element of $\mathscr{R}[X]$ of degree m cannot have more than m distinct roots.*

If $\mathscr{R}$ is a commutative ring then the concepts of right and left divisibility coincide. Moreover, if $\mathscr{R}$ is an integral domain and if p is an element of $\mathscr{R}$ then p is called a *prime* or an *irreducible* element of $\mathscr{R}$ if in any factorization $p = ab$ of p the element a or b is a nonsingular (*see* page 9) element of $\mathscr{R}$. Otherwise, p is called a *reducible* element of $\mathscr{R}$.

Two elements r and s of an integral domain $\mathscr{R}$ are called *relatively prime* or *prime to each other* if every common divisor of r and s is a nonsingular element of $\mathscr{R}$.

An integral domain in which every nonzero element is uniquely expressible as a (finite) product of primes, except for nonsingular factors and the order of the factors, is called a *unique factorization domain.*

Thus, the set of all integers with their usual addition and multiplication is a unique factorization domain. Also, every field is a unique factorization domain. Moreover, it can be shown that

(vi) $\mathscr{R}[X]$ is a unique factorization domain if and only if $\mathscr{R}$ is a unique factorization domain.

The concept of a greatest common divisor can be introduced in any commutative ring.

Let $\mathscr{R}$ be a commutative ring and a and b elements of $\mathscr{R}$. Then an element g of $\mathscr{R}$ is called a *greatest common divisor* (briefly g.c.d) of a and b if g divides both a and b, and, if every divisor of both a and b divides g.

It can be readily verified that in a unique factorization domain $\mathscr{R}$ every two elements have a g.c.d. Moreover, two elements a and b of $\mathscr{R}$ are relatively prime if a nonsingular element of $\mathscr{R}$ is a g.c.d of a and b.

Another example of a commutative ring in which every two elements have a g.c.d is the so-called principal ideal ring.

An integral domain $\mathscr{R}$ is called a *principal ideal ring* if every ideal of $\mathscr{R}$ is a principal ideal (*see* page 11). Thus, the set of all integers with their usual addition and multplication is a principal ideal ring. Also, every field is a principal ideal ring.

It can be shown that a principal ideal ring is a unique factorization domain. However, the converse is not true in general.

If a and b are elements of a principal ideal ring $\mathscr{R}$ then g is a g.c.d of a and b if the ideal generated by g is equal to the ideal generated by $\{a, b\}$. Clearly, in this case we have

$$g = ra + sb \qquad (46)$$

for some elements r and s of $\mathscr{R}$.

In connection with (46) let us mention that two elements a and b of an integral domain $\mathscr{R}$ may not have a g.c.d. However, if g is a g.c.d of a and b in $\mathscr{R}$ and if g is expressible in the form as given by (46) then g is a g.c.d of a and b in any integral domain $\mathscr{R}'$ such that $\mathscr{R} \subset \mathscr{R}'$.

Although in a principal ideal ring every two elements have a g.c.d, there is no algorithm for actually computing it, in general.

On the other hand, if $\mathscr{F}$ is a field then in the integral domain $\mathscr{F}[X]$ the Euclidean algorithm mentioned above provides the familiar method for actually computing a g.c.d of any two elements of $\mathscr{F}[X]$ which are not both zero. Moreover, as a consequence of the Euclidean algorithm, for every two elements P and Q of $\mathscr{F}[X]$ which are not both zero there exists a unique *monic* element G of $\mathscr{F}[X]$ such that G divides both P and Q and

$$G = AP + BQ \tag{47}$$

for some elements A and B of $\mathscr{F}[X]$. Clearly, it is natural to call G *the greatest common divisor* of P and Q. If P and Q are both 0 then 0 is their only g.c.d and hence in this case 0 is the g.c.d of P and Q.

Without giving a definition, we mention here that there are commutative rings in which it is possible to have a division algorithm similar to that of Euclidean. These rings are called *Euclidean rings*. In a Euclidean ring a g.c.d of every two elements is expressible in the form (46) and is actually computable.

The set of all integers with their usual addition and multiplication is a Euclidean ring. Also, every field is a Euclidean ring. Moreover, if $\mathscr{F}$ is a field then in view of (47) the integral domain $\mathscr{F}[X]$ is a Euclidean ring.

It can be shown that every Euclidean ring is a principal ideal ring. However, the converse is not true, in general.

As observed in the case of (46), in view of (47) we make the following observations. Let $\mathscr{F}$ be a field and G be the g.c.d of two elements P and Q of $\mathscr{F}[X]$. Then G is the g.c.d of P and Q in $\mathscr{F}'[X]$ where $\mathscr{F}'$ is any field which contains $\mathscr{F}$ as a subfield. Consequently, if P and Q are relatively prime in $\mathscr{F}[X]$ they remain relatively prime in $\mathscr{F}'[X]$. On the other hand, however, if P is an irreducible element of $\mathscr{F}[X]$ it may happen that P becomes a reducible element of $\mathscr{F}'[X]$.

In closing this section we mention that the concept of the polynomial ring $\mathscr{R}[X]$ in one indeterminate X over the ring $\mathscr{R}$ can be generalized to that of a polynomial ring $\mathscr{R}[X_1, X_2, \ldots, X_n]$ in several indeterminates $X_1, X_2, \ldots, X_n$ over the ring $\mathscr{R}$. Denoting $\mathscr{R}[X_1, X_2, \ldots, X_n]$ by $\mathscr{R}_n$ we define $\mathscr{R}[X_1, X_2, \ldots, X_n]$ inductively as follows:

$$\mathscr{R}_1 = \mathscr{R}[X_1], \mathscr{R}_2 = \mathscr{R}_1[X_2], \ldots, \mathscr{R}_n = \mathscr{R}_{n-1}[X_n].$$

We remark only that many properties of $\mathscr{R}[X]$ generalize in an obvious way to those of $\mathscr{R}[X_1, X_2, \ldots, X_n]$.

Exercises

1. Let P be a nonzero element of $\mathscr{R}[X]$ where $\mathscr{R}$ is a commutative ring. Prove that P is a divisor of zero in $\mathscr{R}[X]$ if and only if there exists a nonzero element r of $\mathscr{R}$ such that $rP = 0$.

2. Let $\mathscr{R}$ be a commutative ring with unity. Prove that $a_n X^n + \cdots + a_1 X + a_0$ is a nonsingular element of $\mathscr{R}[X]$ if and only if a_0 is a nonsingular element of $\mathscr{R}$ and $a_1, a_2, \ldots, a_n$ are nilpotent elements of $\mathscr{R}$.

3. Let $\mathscr{R}$ be an integral domain. Prove that P is a nonsingular element of $\mathscr{R}[X]$ if and only if P is a nonsingular element of $\mathscr{R}$.

4. Let $\mathscr{R}$ be an integral domain. Prove that an irreducible element P of $\mathscr{R}[X]$ is relatively prime to every element of $\mathscr{R}[X]$ whose degree is less than that of P.

5. Let $\mathscr{F}$ be a field and P be an element of degree 2 or 3 of $\mathscr{F}[X]$. Prove that P is a reducible element of $\mathscr{F}[X]$ if and only if $P(c) = 0$ for some element c of $\mathscr{F}$.

6. Let $\mathscr{F}$ be a field. Prove that $X^4 + X^3 + X + 1$ is a reducible element of $\mathscr{F}[X]$.

7. Let $p > 1$ be a prime number. Prove that $X^n - p$ as well as $X^{p-1} + \cdots + X + 1$ is an irreducible element of $\mathscr{F}[X]$ where $\mathscr{F}$ is the field of all rational numbers.

8. Let $a_0, a_1, \ldots, a_n$ be integers and p be a prime number such that p does not divide a_n, p^2 does not divide a_0 and p divides $a_1, a_2, \ldots$ and a_{n-1}. Prove that $a_n X^n + \cdots + a_1 X + a_0$ is an irreducible element of $\mathscr{F}[X]$ where $\mathscr{F}$ is the field of all rational numbers.

9. Prove that $\mathscr{R}$ is an integral domain if and only if $\mathscr{R}[X_1, X_2, \ldots, X_n]$ is an integral domain.

10. Prove that $\mathscr{R}$ is a unique factorization domain if and only if $\mathscr{R}[X_1, X_2, \ldots, X_n]$ is a unique factorization domain.

11. Let $\mathscr{F}$ be a field. Prove that $\mathscr{F}[X_1, X_2]$ is not a principal ideal ring.

12. Let $\mathscr{R}$ be an integral domain which is not a field. Prove that $\mathscr{R}[X]$ is not a principal ideal ring.

13. Prove that if a singular element p of a principal ideal ring $\mathscr{R}$ is a prime element of $\mathscr{R}$ then $\mathscr{R}/(p)$ is a field where (p) is the ideal generated by p.

14. Let Q be the ring of real quaternions. Prove that there are infinitely many elements q of Q such that $q^2 + 1 = 0$.

15. Let $\mathscr{F}$ be a field. Prove that a nonzero ideal of $\mathscr{F}[X]$ is a prime ideal of $\mathscr{F}[X]$ if and only if it is generated by a monic irreducible element of $\mathscr{F}[X]$. Prove also that every nontrivial prime ideal of $\mathscr{F}[X]$ is a maximal ideal of $\mathscr{F}[X]$.

16. Prove that in a finite field with n elements every element is a root of the polynomial $X^n - X$.

17. Let $\mathscr{F}$ be a finite field with n elements and p be an irreducible element of degree m of $\mathscr{F}[X]$. Prove that P divides $X^{n^e} - X$ if and only if m divides e where e is a natural number.

18. Let $\mathscr{R}$ be a commutative ring without divisors of zero. Prove that an element p of $\mathscr{R}$ is a prime element if and only if the quotient ring $\mathscr{R}/(p)$ is a commutative ring without divisors of zero where (p) is the ideal generated by p.

19. Let $\mathscr{F}$ be a finite field with n elements. Using the usual definition of the degree of a polynomial in two indeterminates prove that if $P(X_1, X_2)$ is a nonzero polynomial of degree less than n over $\mathscr{F}$ then there exist elements c_1 and c_2 of $\mathscr{F}$ such that $P(c_1, c_2) \neq 0$.

20. Let $\mathscr{F}$ be a finite field with n elements and let $P(X_1, X_2, X_3)$ be a polynomial of degree less than 3 over $\mathscr{F}$. Prove that if $P(0, 0, 0) = 0$ then there exist elements c_1, c_2 and c_3 of $\mathscr{F}$ not all zero such that $P(c_1, c_2, c_3) = 0$.

1.5. Matrix Rings

Another method of forming new rings based on the structure of a given ring consists of forming matrix rings over a given ring $\mathscr{R}$ (*i.e.*, with entries in $\mathscr{R}$). Here again we denote the elements of $\mathscr{R}$ by small letters. In particular, we denote the unity of $\mathscr{R}$ by 1, if it exists.

As usual, by an *m by n matrix* (where m and n are positive natural numbers) over a ring $\mathscr{R}$ we mean a rectangular array

$$A = \begin{pmatrix} a_{11} & a_{12} & \cdots & a_{1n} \\ a_{21} & a_{22} & \cdots & a_{2n} \\ \cdots & \cdots & \cdots & \cdots \\ a_{m1} & a_{m2} & \cdots & a_{mn} \end{pmatrix}$$

with *m rows* and *n columns* where the entries a_{ij} are elements of $\mathscr{R}$. It is customary to abbreviate the above as

$$A = (a_{ij}), \qquad (i = 1, 2, \ldots, n, \quad j = 1, 2, \ldots, m)$$

and when no confusion is likely to arise, simply as

$$A = (a_{ij}).$$

As expected, we define equality between an m by n matrix (a_{ij}) and a p by q matrix (b_{ij}) by requiring that their respective dimensions and entries be equal, *i.e.*,

$$(a_{ij}) = (b_{ij}) \qquad \text{if } m = p, n = q \text{ and } a_{ij} = b_{ij}. \tag{48}$$

We define the *sum $A + B$* of two m by n matrices $A = (a_{ij})$ and $B = (b_{ij})$ as being the m by n matrix $S = (s_{ij})$ where

$$s_{ij} = a_{ij} + b_{ij}. \tag{49}$$

We define the *product AB* of an m by n matrix $A = (a_{ij})$ and an n by q matrix $B = (b_{jk})$ as being the m by q matrix $P = (p_{ik})$ where

$$p_{ik} = \sum_{j=1}^{n} a_{ij} b_{jk}. \tag{50}$$

Let us denote the set of all m by m matrices over a ring $\mathscr{R}$ by $\mathscr{R}_m$. In view of (48) it can be readily verified that $(\mathscr{R}_m, +, \cdot)$ with addition and multiplication among the elements of $\mathscr{R}_m$ defined respectively according to (49) and (50) is a ring.

The ring $(\mathscr{R}_m, +, \cdot)$ is called the *total m by m matrix ring over the ring* $\mathscr{R}$ (naturally, the entries of the elements of $\mathscr{R}_m$ are elements of $\mathscr{R}$).

Denoting the zero of $\mathscr{R}$ by 0, it is clear that the zero of $\mathscr{R}_m$ is the matrix

29

$0_m = (a_{ij})$ where $a_{ij} = 0$ for $i, j = 1, 2, \ldots, m$. Moreover, $\mathcal{R}_m$ has a unity if and only if $\mathcal{R}$ has a unity 1, in which case, the unity of $\mathcal{R}_m$ is the matrix

$$I_m = (a_{ij}) \qquad \text{with} \qquad a_{ij} = \delta_j{}^i \qquad (i, j = 1, 2, \ldots, m)$$

where $\delta_j{}^i$ is the familiar Kronecker delta, *i.e.*, $\delta_j{}^i = 0$ if $i \neq j$ and $\delta_j{}^i = 1$ if $i = j$.

When no confusion is likely to arise 0_m is denoted by 0 and I_m by I, or even by 1.

Clearly, for $m = 1$ the total matrix ring $\mathcal{R}_m$ over the ring $\mathcal{R}$ is isomorphic to the ring $\mathcal{R}$, *i.e.*, $\mathcal{R}_1 \cong \mathcal{R}$.

As shown below, for $m > 1$ the ring $\mathcal{R}_m$ may acquire some properties which are not shared by $\mathcal{R}$. However, this does not preclude $\mathcal{R}_m$ from inheriting some of the specific properties of $\mathcal{R}$.

If a ring $\mathcal{R}$ has at least one nonzero element r then for $m > 1$ the ring $\mathcal{R}_m$ has divisors of zero. This can be seen from the following example.

$$\begin{pmatrix} 0 & 0 \\ 0 & r \end{pmatrix} \begin{pmatrix} 0 & r \\ 0 & 0 \end{pmatrix} = \begin{pmatrix} 0 & 0 \\ 0 & 0 \end{pmatrix}. \tag{51}$$

For $m > 1$, even if $\mathcal{R}_m$ is a commutative ring then only in very exceptional cases is $\mathcal{R}_m$ commutative. In general $\mathcal{R}_m$ is a noncommutative ring. This can be seen from comparing (51) with

$$\begin{pmatrix} 0 & r \\ 0 & 0 \end{pmatrix} \begin{pmatrix} 0 & 0 \\ 0 & r \end{pmatrix} = \begin{pmatrix} 0 & r^2 \\ 0 & 0 \end{pmatrix}. \tag{52}$$

If $m > 1$, then, in general, $\mathcal{R}_m$ always has nonzero nilpotent elements as shown by

$$\begin{pmatrix} 0 & r \\ 0 & 0 \end{pmatrix} \begin{pmatrix} 0 & r \\ 0 & 0 \end{pmatrix} = \begin{pmatrix} 0 & 0 \\ 0 & 0 \end{pmatrix}. \tag{53}$$

If e is an idempotent element of $\mathcal{R}$ then each of

$$\begin{pmatrix} e & 0 \\ 0 & e \end{pmatrix}, \qquad \begin{pmatrix} e & 0 \\ 0 & 0 \end{pmatrix}, \qquad \begin{pmatrix} 0 & 0 \\ 0 & e \end{pmatrix} \tag{54}$$

is an idempotent element of $\mathcal{R}_2$.

In particular, if $\mathcal{R}$ has a unity element then as (51), (52), (53) and (54) show for $m > 1$ the ring $\mathcal{R}_m$ has divisors of zero, is not commutative, has nonzero nilpotent elements and has nonzero idempotent elements which are not necessarily equal to the unity of $\mathcal{R}_m$.

For $m > 1$, the total matrix ring $\mathcal{R}_m$ has many subrings. For instance, the set of all m by m *triangular* matrices (*i.e.*, a square matrix (a_{ij}) such that $a_{ij} = 0$ for $i > j$ or $a_{ij} = 0$ for $j > i$) over $\mathcal{R}$ is a subring of $\mathcal{R}_m$. Moreover, the set of all m by m *diagonal* matrices (*i.e.*, a square matrix (a_{ij})

such that $a_{ij} = 0$ for $i \neq j$) over $\mathcal{R}$ is a subring of $\mathcal{R}_m$. Furthermore, the set of all m by m *scalar* matrices (*i.e.*, a diagonal matrix (a_{ij}) such that $a_{ii} = r$ for some $r \in \mathcal{R}$) is a subring of $\mathcal{R}_m$ which is isomorphic to the ring $\mathcal{R}$.

As mentioned above, the total matrix ring $\mathcal{R}_m$ is noncommutative in general. In this connection, however, we have

LEMMA 8. *Let $\mathscr{C}$ be the center of a ring $\mathcal{R}$ with unity 1. Then the center $\mathscr{C}_{\overline{m}}$ of the total matrix ring $\mathcal{R}_m$ is the set $\mathscr{S}$ of all m by m scalar matrices (a_{ij}) with $a_{ii} \in \mathscr{C}$.*

Proof. Clearly, $\mathscr{S} \subset \mathscr{C}_{\overline{m}}$. Thus, it is enough to prove that if $A \in \mathscr{C}_m$ then $A \in \mathscr{S}$. We prove the Lemma for the case $m = 2$ which illustrates the proof in general. Let

$$A = \begin{pmatrix} x & y \\ z & u \end{pmatrix}.$$

If $A \in \mathscr{C}_{\overline{m}}$ then it must commute with every element of $\mathcal{R}_m$. Hence, in particular

$$\begin{pmatrix} 1 & 0 \\ 0 & 0 \end{pmatrix}\begin{pmatrix} x & y \\ z & u \end{pmatrix} = \begin{pmatrix} x & y \\ 0 & 0 \end{pmatrix} = \begin{pmatrix} x & 0 \\ z & 0 \end{pmatrix} = \begin{pmatrix} x & y \\ z & u \end{pmatrix}\begin{pmatrix} 1 & 0 \\ 0 & 0 \end{pmatrix}$$

implying that $y = z = 0$. Moreover,

$$\begin{pmatrix} 0 & 1 \\ 0 & 0 \end{pmatrix}\begin{pmatrix} x & 0 \\ 0 & u \end{pmatrix} = \begin{pmatrix} 0 & u \\ 0 & 0 \end{pmatrix} = \begin{pmatrix} 0 & x \\ 0 & 0 \end{pmatrix} = \begin{pmatrix} x & 0 \\ 0 & u \end{pmatrix}\begin{pmatrix} 0 & 1 \\ 0 & 0 \end{pmatrix}$$

implying that $x = u$. Therefore, A is the scalar matrix

$$A = \begin{pmatrix} x & 0 \\ 0 & x \end{pmatrix}.$$

But since $A \in \mathscr{C}_{\overline{m}}$ it is readily seen that x must be an element of $\mathscr{C}$, as desired.

So far as the ideals of the total matrix ring $\mathcal{R}_m$ are concerned, let us observe that if $\mathscr{I}$ is an ideal of $\mathcal{R}$ then obviously $\mathscr{I}_m$ is an ideal of $\mathcal{R}_m$. On the other hand, we have:

LEMMA 9. *Let $\mathscr{S}$ be an ideal of a total matrix ring $\mathcal{R}_m$ where $\mathcal{R}$ is a ring with unity 1. Then the set $\mathscr{I}$ of all entries of all the elements of $\mathscr{S}$ is an ideal of $\mathcal{R}$.*

Proof. To prove the lemma it is enough to show that if x and y are elements of $\mathscr{I}$ then so are $x - y$, xr and rx for every element r of $\mathcal{R}$. Again, we give the proof for the case $m = 2$. Now, let

$$\begin{pmatrix} p & q \\ x & t \end{pmatrix} \quad \text{and} \quad \begin{pmatrix} u & y \\ v & w \end{pmatrix}$$

be elements of the ideal $\mathscr{S}$. But then

$$\begin{pmatrix} p & q \\ x & t \end{pmatrix}\begin{pmatrix} 0 & 1 \\ 0 & 0 \end{pmatrix} - \begin{pmatrix} 0 & 0 \\ 1 & 0 \end{pmatrix}\begin{pmatrix} u & y \\ v & w \end{pmatrix} = \begin{pmatrix} 0 & p \\ -u & x-y \end{pmatrix}$$

is an element of $\mathscr{S}$ implying that $x-y$ is an element of $\mathscr{I}$. Similarly, since $\mathscr{S}$ is an ideal

$$\begin{pmatrix} p & q \\ x & t \end{pmatrix}\begin{pmatrix} 0 & r \\ 0 & 0 \end{pmatrix} = \begin{pmatrix} 0 & pr \\ 0 & xr \end{pmatrix} \qquad \text{and} \qquad \begin{pmatrix} 0 & r \\ 0 & 0 \end{pmatrix}\begin{pmatrix} p & q \\ x & t \end{pmatrix} = \begin{pmatrix} rx & rt \\ 0 & 0 \end{pmatrix}$$

are elements of $\mathscr{S}$ implying that xr as well as rx is an element of $\mathscr{I}$ for every element r of $\mathscr{R}$.

COROLLARY 1. *If $\mathscr{R}$ is a simple ring with unity (in particular a division ring or a field) then the total matrix ring $\mathscr{R}_m$ is also a simple ring with unity.*

We assume that the reader is acquainted with the notions of the *determinant* ($\det M$) of a square matrix M, the *minor* and the *cofactor* of an entry of M, the *adjoint matrix* ($\operatorname{adj} M$) of M and with the fundamental equality

$$M(\operatorname{adj} M) = (\operatorname{adj} M)M = D \tag{55}$$

where M is an m by m matrix over a commutative ring and D is an m by m scalar matrix on whose diagonal every entry is equal to the determinant of M.

As a consequence of (55), we have

COROLLARY 2. *If $\mathscr{R}$ is a commutative ring then an element A of $\mathscr{R}_m$ is nonsingular (or a divisor of zero) if and only if $\det A$ is a nonsingular element (or zero or a divisor of zero) of $\mathscr{R}$.*

Next, we consider the set $\mathscr{R}_{mn}$ of all m by n matrices over a ring $\mathscr{R}$.

Clearly, $(\mathscr{R}_{mn}, +)$, i.e., $\mathscr{R}_{mn}$ under matrix addition, is an abelian group. Moreover, if P and Q are m by m and n by n nonsingular matrices respectively, then each of the mappings

$$X \to PX, \qquad X \to XQ, \qquad X \to PXQ \tag{56}$$

is an *automorphism* of $(\mathscr{R}_{mn}, +)$ i.e., an isomorphism from $(\mathscr{R}_{mn}, +)$ onto $(\mathscr{R}_{mn}, +)$.

Let us observe that each of the mappings in (56) is obtained by premultiplying or postmultiplying X by a nonsingular m by m matrix P or a nonsingular n by n matrix Q, respectively.

In this connection we recall the following three types of *elementary row* (or *column*) *transformations* of an m by n matrix M over a ring $\mathscr{R}$.

(i) *Interchanging the i-th and the j-th rows (or columns) of M.*

Transformation (*i*) amounts to premultiplying (or postmultiplying) *M* by an *m* by *m* (or *n* by *n*) matrix L_1 (or R_1) which results from interchanging the *i*-th and the *j*-th rows (or columns) of the *m* by *m* (or *n* by *n*) unit matrix I_m (or I_n) over $\mathscr{R}$.

(ii) *Adding to the i-th row (or column) of M the multiple by an element r of $\mathscr{R}$ of the j-th row (or column) of M where $i \neq j$.*

Transformation (ii) amounts to premultiplying (or postmultiplying) *M* by an *m* by *m* (or *n* by *n*) matrix L_2 (or R_2) which results from adding to the *i*-th row (or column) of I_m (or I_n) the multiple by *r* of the *j*-th row (or column) of I_m (or I_n) where $i \neq j$.

(iii) *Multiplying the i-th row (or column) of M by a nonsingular element a of $\mathscr{R}$.*

Transformation (iii) amounts to premultiplying (or postmultiplying) *M* by an *m* by *m* (or *n* by *n*) matrix L_3 (or R_3) which results from multiplying the *i* th row (or column) of I_m (or I_n) by *a*.

Clearly, any matrix of type L_1, L_2 and L_3 (or R_1, R_2 and R_3) as well as any finite product of matrices of these types is a nonsingular matrix.

Returning to the automorphism of type

$$X \rightarrow PXQ \tag{57}$$

of $(\mathscr{R}_{mn}, +)$ where *P* and *Q* are *m* by *m* and *n* by *n* nonsingular matrices, respectively, it is natural to ask if there exists an automorphism of type (57) such that the image *PMQ* of a given *m* by *n* matrix *M* is a diagonal *m* by *n* matrix (*i.e.*, an *m* by *n* matrix (a_{ij}) such that $a_{ij} = 0$ for $i \neq j$).

We give a positive answer to the above question for the two particular cases where $\mathscr{R}$ is the ring of all integers or where $\mathscr{R}$ is the ring of all polynomials $\mathscr{F}[x]$ over a field $\mathscr{F}$.

In what follows we say that two *m* by *n* matrices *A* and *B* over a ring $\mathscr{R}$ are *rationally equivalent* and we denote this by:

$$A \simeq B \qquad \text{if } A = PBQ \tag{58}$$

where *P* and *Q* are nonsingular matrices over $\mathscr{R}$. Clearly, "rational equivalence" is an equivalence relation in $\mathscr{R}_{mn}$.

LEMMA 10. *Let M be an m by n matrix with integer entries. Then there exist nonsingular matrices P and Q with integer entries such that*

$$M = P \begin{pmatrix} g_1 & 0 & 0 & \cdot & 0 \\ 0 & g_2 & 0 & \cdot & 0 \\ 0 & 0 & g_3 & \cdot & 0 \\ \cdot & \cdot & \cdot & \cdot & \cdot \\ 0 & 0 & 0 & \cdot & \cdot \end{pmatrix} Q \tag{59}$$

where $\{g_1, g_2, g_3, \ldots, g_k\}$ *with* $k = \min\{m, n\}$ *is a unique set of non-negative integers such that* g_1 *divides* g_2, g_2 *divides* $g_3, \ldots, g_{k-1}$ *divides* g_k.

Proof. If M is the zero m by n matrix then to prove the lemma it is enough to take $g_i = 0$. Thus, in what follows let M be a nonzero m by n matrix and let $\mathscr{E}$ be the set of all m by n matrices with integer entries which are rationally equivalent to M. Let g_1 be the smallest positive absolute value of all the entries of all the elements of $\mathscr{E}$. In view of transformations of type (i) and (iii) there exists an element $H = (h_{ij})$ of $\mathscr{E}$ such that $h_{11} = g_1$. By the Euclidean division algorithm for integers, we may write j_{1i} uniquely as $g_1 q_i + r_i$ with $0 \leq r_i < g_1$. However, in view of transformation of type (ii), if we add the multiple by $-q_i$ of the first column of H to its i-th column, we would arrive at a contradiction, unless $r_i = 0$. With a similar reasoning for the case of h_{i1} we conclude that M is rationally equivalent to a matrix (e_{ij}) such that $e_{11} = g_1$ and $e_{1i} = e_{i1} = 0$ if $i \neq 1$. After finitely many such steps we see that M is rationally equivalent to an m by n matrix (g_{ij}) such that $g_{ij} = 0$ if $i \neq j$ and $g_{11} = g_1$. Moreover, $g_{ii} = g_i$ for $i > 1$ where g_i is the smallest absolute value (priority given to the smallest positive absolute value if it exists) of all the entries of all the matrices which are rationally equivalent to the $m+1-i$ by $n+1-i$ matrix (r_{ij}) with $r_{ij} = 0$ if $i \neq j$ and $r_{11} = g_i, r_{22} = g_{i+1}, \ldots, r_{kk} = g_k$ where $k = \min\{m, n\}$.

Next, let us examine

$$\begin{pmatrix} g_i & 0 & \cdot \\ 0 & g_{i+1} & \cdot \\ \cdot & \cdot & \cdot \end{pmatrix}.$$

By the Euclidean division algorithm for integers we have $g_{i+1} = g_i u_i + v_i$ with $0 \leq v_i < g_i$. But then in view of transformations of type (ii) we may obtain

$$\begin{pmatrix} g_i & -g_i u_i & \cdot \\ g_i & v_i & \cdot \\ \cdot & \cdot & \cdot \end{pmatrix}$$

which would lead to a contradiction unless $v_i = 0$, *i.e.*, g_i divides g_{i+1}, as desired.

Thus, M is rationally equivalent to a matrix described in (59). The uniqueness of the g_i's follows from their construction.

In (59) each of the (nonnegative integers) $g_1, g_2, \ldots, g_k$ is called an *invariant factor* of the matrix M.

It is customary to relabel the invariant factors of a matrix according to the nonincreasing order of their magnitudes. Thus, if $g_1, g_2, \ldots, g_r$ are the r nonzero diagonal entries in (59) then g_r is denoted by f_1 and is called the *first nonzero invariant factor* of M. Similarly, g_{r-1} is denoted by f_2 and is called the *second nonzero invariant factor* of M, and so forth.

In view of Lemma 10 and transformations of type (i) we have

LEMMA 11. *Every m by n matrix M with integer entries is rationally equivalent to an m by n matrix (a_{ij}) with integer entries such that $a_{ij} = 0$ for $i \neq j$ and $a_{ii} = f_i$ where f_i is the i-th nonzero invariant factor of M or $a_{ii} = 0$.*

Clearly, if $f_1, f_2, \ldots, f_r$ are the r nonzero invariant factors of M then f_{i+1} divides f_i for $1 \leq i \leq r-1$.

In terms of the invariant factors, Lemma 10 yields:

COROLLARY 3. *Two m by n matrices over the ring of integers are rationally equivalent if and only if they have the same invariant factors.*

An invariant factor of a matrix M (over the ring of integers) is called a *trivial elementary divisor* of M if it is equal to 0 or 1. Otherwise, by definition an *elementary divisor* of M is a power of a prime p^s (with $p > 1$ and $s > 0$) such that p^s enters in the unique factorization (into prime power factors) of an invariant factor of M.

Thus, if a matrix over the ring of integers is such that its invariant factors are:

$$24, \quad 12, \quad 1, \quad 1, \quad 0$$

then its elementary divisors are:

$$3, \quad 3, \quad 2^3, \quad 2^2, \quad 1, \quad 1, \quad 0.$$

Conversely, if a matrix over the ring of integers is such that its elementary divisors are:

$$3^2, \quad 3, \quad 3, \quad 2^3, \quad 2^2, \quad 1, \quad 1, \quad 0, \quad 0$$

then its invariant factors are:

$$3^2 \cdot 2^3 = 72, \quad 3 \cdot 2^2 = 12, \quad 3, \quad 1, \quad 1, \quad 0, \quad 0$$

In view of the above and Corollary 3 we have

COROLLARY 4. *Two m by n matrices over the ring of integers are rationally equivalent if and only if they have the same elementary divisors.*

As an example, let us observe that over the ring of integers

$$\begin{pmatrix} 1 & 1 \\ 0 & 1 \end{pmatrix} \simeq \begin{pmatrix} 1 & 0 \\ 0 & 1 \end{pmatrix} \tag{60}$$

because

$$\begin{pmatrix} 1 & -1 \\ 0 & 1 \end{pmatrix}\begin{pmatrix} 1 & 1 \\ 0 & 1 \end{pmatrix}\begin{pmatrix} 1 & 0 \\ 0 & 1 \end{pmatrix} = \begin{pmatrix} 1 & 0 \\ 0 & 1 \end{pmatrix}.$$

The invariant factors of either matrix in (60) are 1 and 1 and so are their elementary divisors.

Since the proof of Lemma 10 is primarily based on the Euclidean division algorithm and since Euclidean division algorithm (*see* page 25) is valid in the polynomial ring $\mathscr{F}[x]$ where $\mathscr{F}$ is a field, according to Lemma 11 we have:

LEMMA 12. *Let M be an m by n matrix over the polynomial ring $\mathscr{F}[x]$ where $\mathscr{F}$ is a field. Then there exist nonsingular matrices P and Q over $\mathscr{F}[x]$ such that*

$$M = P \begin{pmatrix} f_1 & 0 & 0 & \cdot & 0 \\ 0 & f_2 & 0 & \cdot & 0 \\ 0 & 0 & f_3 & \cdot & 0 \\ \cdot & \cdot & \cdot & \cdot & \cdot \\ 0 & 0 & 0 & \cdot & \cdot \end{pmatrix} Q \tag{61}$$

where $\{f_1, f_2, f_3, \ldots, f_k\}$ with $k = \min\{m, n\}$ is a unique set of monic polynomials (including the unity polynomial 1) over $\mathscr{F}$ or 0 and such that if there are r nonzero diagonal entries in (61) then f_2 divides f_1, f_3 divides f_2 and f_r divides f_{r-1}.

Naturally, in the above, $f_1, f_2, \ldots, f_k$ are called the first, the second, ..., the k-th invariant factor of the matrix M.

Let us observe that since x has no multiplicative inverse in $\mathscr{F}[x]$, and since P and Q are nonsingular matrices over $\mathscr{F}[x]$, their determinants are nonzero elements of $\mathscr{F}$ and are independent of x.

The notions of rational equivalence and the elementary divisor for the case of m by n matrices over the polynomial ring $\mathscr{F}[x]$ are introduced in the obvious manner.

Thus, following Corollary 3 we have:

COROLLARY 5. *Two m by n matrices over the polynomial ring $\mathscr{F}[x]$ where $\mathscr{F}$ is a field, are rationally equivalent if and only if they have the same invariant factors.*

The passage from the invariant factors to the elementary divisors (and vice versa) of a matrix over $\mathscr{F}[x]$ where $\mathscr{F}$ is a field, is accomplished as in the case of a matrix over the ring of integers.

Thus, if a matrix over the polynomial ring $\mathscr{F}[x]$ where $\mathscr{F}$ is the field of real numbers, is such that its invariant factors are:

$$(x^2+1)^2(x-5)^3, \quad (x^2+1)(x-5)^2, \quad 1, \quad 1, \quad 0$$

then its elementary divisors are:

$$(x^2+1)^2, \quad (x^2+1), \quad (x-5)^3, \quad (x-5)^2, \quad 1, \quad 1, \quad 0.$$

Conversely, if a matrix over $\mathscr{F}[x]$ where $\mathscr{F}$ is the field of real numbers, is such that its elementary divisors are:

$$(x^2+1)^5, \quad (x^2+2)^3, \quad (x^2+2), \quad (x-1)^3, \quad 1, \quad 0, \quad 0$$

then its invariant factors are:

$$(x^2+1)^5(x^2+2)^3(x-1)^3, \quad (x^2+2), \quad 1, \quad 0, \quad 0.$$

Following Corollary 4 we have

COROLLARY 6.   *Two m by n matrices over the polynomial ring $\mathscr{F}[x]$ where $\mathscr{F}$ is a field, are rationally equivalent if and only if they have the same elementary divisors.*

As an example, let us observe that over $\mathscr{F}[x]$ where $\mathscr{F}$ is the field of real numbers

$$\begin{pmatrix} 1-x & 1 \\ 0 & 1-x \end{pmatrix} \simeq \begin{pmatrix} (1-x)^2 & 0 \\ 0 & 1 \end{pmatrix} \qquad (62)$$

because

$$\begin{pmatrix} 1-x & -1 \\ 1 & 0 \end{pmatrix} \begin{pmatrix} 1-x & 1 \\ 0 & 1-x \end{pmatrix} \begin{pmatrix} 1 & 0 \\ x-1 & 1 \end{pmatrix} = \begin{pmatrix} (1-x)^2 & 0 \\ 0 & 1 \end{pmatrix}.$$

The invariant factors of either matrix in (62) are $(1-x)^2$ and 1 and so are their elementary divisors.

Exercises

1. Let $\mathscr{D}$ be a division ring. Prove that an element of the total m by m matrix ring $\mathscr{D}_m$ is nonsingular if and only if it is not a left (or right) divisor of zero.

2. Let $\mathscr{D}$ be a division ring. Prove that an element of the total m by m matrix ring $\mathscr{D}_m$ is nonsingular if and only if it is a product of matrices which are obtained from the unity matrix I_m via transformations of type (i), (ii) or (iii) mentioned on page 33.

3. Find the invariant factors and the elementary divisors (over the ring of integers) of a 3 by 3 matrix (a_{ij}) such that $a_{ij} = 1$ for $i, j = 1, 2, 3$.

4. Let $\mathscr{F}$ be the field of all real numbers. Find the invariant factors and the elementary divisors (over the polynomial ring $\mathscr{F}[x]$) of a 3 by 3 matrix (a_{ij}) such that $a_{ij} = x$ for $i, j = 1, 2, 3$.

5. Let M be an m by n matrix over the ring of all integers. Prove that if M has r nonzero invariant factors f_i then

$$f_i = \frac{d_{r+1-i}}{d_{r-1}} \qquad (i = 1, 2, \ldots, r)$$

where $d_0 = 1$ and d_k is the g.c.d of all k by k minors of M for $k > 0$.

6. Prove the result analogous to that of Problem 5 for the case of matrices over the polynomial ring $\mathscr{F}[x]$ where $\mathscr{F}$ is a field.

7. Let $\mathscr{R}$ be a ring such that $X \in \mathscr{R}X\mathscr{R}$ for every $X \in \mathscr{R}$. Prove that every ideal of the total m by m matrix ring over $\mathscr{R}$ is a total m by m matrix ring over an ideal of $\mathscr{R}$.

8. Let $\mathscr{R}$ be a simple ring such that $\mathscr{R}\mathscr{R} \neq \{0\}$. Prove that every total m by m matrix ring $\mathscr{R}_m$ over $\mathscr{R}$ is simple and that $\mathscr{R}_m\mathscr{R}_m \neq \{0\}$.

9. Let $\mathscr{R}$ be a commutative ring with unity and $\mathscr{R}_m$ the total m by m matrix ring over $\mathscr{R}$. Prove that $XY = I_m$ implies $YX = I_m$ for every element X and Y of $\mathscr{R}_m$ where I_m is the unity of $\mathscr{R}_m$.

10. Let $A = (a_{ij})$ be an m by m skew symmetric matrix (*i.e.*, a square matrix such that $a_{ij} = -a_{ji}$) over the field of real numbers. Prove that if m is an odd number then A is singular. Prove also that if A has an inverse then the inverse is a skew symmetric matrix.

11. Let $\mathscr{F}$ be the field of real numbers and $\mathscr{F}_m$ the total m by m matrix ring over $\mathscr{F}$. Prove that for every element X and Y of $\mathscr{F}_m$ if $XY - YX$ commutes with X then $XY - YX$ is nilpotent.

12. Let X and Y be m by m matrices over the field of real numbers and I_m the m by m unity matrix. Prove that $XY - YX - I_m \neq 0$.

13. Let N be a nilpotent element of a total m by m matrix ring $\mathscr{F}_m$ where $\mathscr{F}$ is a field. Prove that $N + I_m$ (where I_m is the m by m unity matrix) is a nonsingular element of $\mathscr{F}_m$.

14. Let $\mathscr{C}$ be the set of all 2 by 2 matrices (a_{ij}) over the field of real numbers such that

$$a_{11} = a_{22} \qquad \text{and} \qquad a_{12} = -a_{21}.$$

Prove that $\mathscr{C}$ with the usual matrix addition and multiplication is a field and is isomorphic to the field of all complex numbers.

15. Let $\mathscr{D}$ be the set of all 2 by 2 matrices (a_{ij}) over the field of complex numbers such that

$$a_{11} = x + iy, \quad a_{22} = x - iy, \quad a_{12} = u + iv, \quad a_{21} = -u + iv.$$

Prove that $\mathscr{D}$ with the usual addition and multiplication is a division ring and is isomorphic to the division ring of real quaternions.

CHAPTER 2

Vector Spaces

2.1. Vector Spaces

The concept of a vector space is fundamental for that of a Linear Associative Algebra. In fact, as we shall see, a vector space $\mathscr{V}$ becomes a Linear Associative Algebra as soon as multiplication (naturally, satisfying some conditions) is defined among the elements of $\mathscr{V}$. Thus, as expected, we precede the study of Linear Associative Algebras by that of vector spaces.

A vector space is a special case of a more general algebraic discipline called a *module over a ring*.

DEFINITION 1. *Let $\mathfrak{F}$ be a field and $\mathscr{V}$ an abelian group (under $+$) such that for every element s of $\mathfrak{F}$ and X of $\mathscr{V}$ there is a unique element of $\mathscr{V}$ denoted by sX. Then $\mathscr{V}$ is called a vector space over $\mathfrak{F}$ if*

$$s(X+Y) = sX+sY \qquad (1)$$
$$(s+t)X = sX+tX \qquad (2)$$
$$s(tX) \ = (st)X \qquad (3)$$
$$1X = X \qquad (4)$$

for every element s, t of $\mathfrak{F}$ and X, Y of $\mathscr{V}$.

Clearly, $+$ in (1) as well as in the right side of (2) stands for the addition operation in $\mathscr{V}$. On the other hand, $+$ in the left side of (2) stands for the addition operation in $\mathfrak{F}$. Moreover, st in (3) is the product in $\mathfrak{F}$ of the elements s and t of $\mathfrak{F}$, and, 1 in (4) is the unity of $\mathfrak{F}$.

Let $\mathscr{V}$ be a vector space over a field $\mathfrak{F}$. It is customary to call an element X of $\mathscr{V}$ a *vector* and an element s of $\mathfrak{F}$ a *scalar*. In this connection we refer to $\mathfrak{F}$ as the *scalar domain* of $\mathscr{V}$ and we call the vector sX the product of the scalar s by the vector X.

In what follows we shall denote vector spaces and their subsets by script letters; scalar domains by German letters; scalars by small letters, and, vectors by capital letters. Naturally these items shall be denoted also by suitable configurations using the appropriate letters. For instance, if A and B denote vectors and s and t denote scalars then the configuration $sA + tB$ denotes a vector.

There will be no danger of confusion if the symbol 0 is used to denote both the *zero vector* and the *zero scalar* especially since in every vector space

$$s0 = 0X = 0 \tag{5}$$

for every scalar s and every vector X.

Clearly, in (5) the first as well as the third occurrence of 0 denotes the zero vector and the second occurrence of 0 denotes the zero scalar.

For the sake of simplicity we shall refer to a vector space $\mathscr{V}$ over a field $\mathfrak{F}$ as a vector space over $\mathfrak{F}$. Moreover, when the specific properties of the underlying field are immaterial for our purposes we shall refer to $\mathscr{V}$ simply as a *vector space*.

In Definition 1, in a product of a scalar s by a vector X we placed s to the left of X. Thus, it would be more appropriate to consider Definition 1 as the definition of a *left vector space over $\mathfrak{F}$*. Naturally, for the definition of a *right vector space over $\mathfrak{F}$* we may take Definition 1 and instead of placing the scalars to the left of the vectors, place them to the right of the vectors. However, since $\mathfrak{F}$ is a field it is easy to see that the concepts of a left and of a right vector space over $\mathfrak{F}$ coincide if sX and Xs are identified for every $s \in \mathfrak{F}$ and $X \in \mathscr{V}$.

The notion of a vector space can be generalized in various ways. If in Definition 1 it is postulated that the scalar domain $\mathfrak{F}$ is a division ring $\mathfrak{D}$ then the resulting system is called a *left module over the division ring $\mathfrak{D}$*. If it is postulated that the scalar domain is a ring $\mathfrak{S}$ with unity then the resulting system is called a *unitary left module over the ring $\mathfrak{S}$*. Finally, if in Definition 1 it is postulated that the scalar domain is merely a ring $\mathfrak{A}$ and if condition (4) is also dropped then the resulting system is called a *left module over the ring $\mathfrak{A}$*.

Examples of vector spaces are numerous. For instance, every field is a vector space over itself. Similarly, the set of all real numbers is a vector space over the field of rational numbers. Likewise, the set of all complex numbers as well as real quaternions is a vector space over the field of real numbers.

Let $\mathscr{V}$ be a vector space over a field $\mathfrak{F}$. Then

$$(-s)X = -(sX) \tag{6}$$

and

$$sX = 0 \quad \text{implies } s = 0 \quad \text{or} \quad X = 0 \tag{7}$$

for every $s \in \mathfrak{F}$ and $X \in \mathscr{V}$.

In (6) the scalar $-s$ is the additive inverse of the scalar s and the vector $-(sX)$ is the additive inverse of the vector sX.

To prove (6) it is enough to observe that by (2) and (5) we have

$$(-s+s)X = 0 = (-s)X + sX$$

therefore $(-s)X = -(sX)$.

To prove (7) it is enough to observe that if $sX = 0$ and $s \neq 0$ then by (5), (3) and (4) we have

$$s^{-1}(sX) = 0 = 1X = X.$$

Thus, $X = 0$, as desired.

In view of (3) and (6) we shall denote either of $s(tX)$ and $(st)X$ simply by stX and either of $(-s)X$ and $-(sX)$ simply by $-sX$.

Let $\mathscr{V}$ be a vector space over a field $\mathfrak{F}$. A subset $\mathscr{S}$ of $\mathscr{V}$ is called a *subspace* of $\mathscr{V}$ if $\mathscr{S}$ is a vector space over $\mathfrak{F}$ (naturally, with respect to the operations in $\mathscr{V}$).

Clearly, a nonempty subset $\mathscr{S}$ of a vector space $\mathscr{V}$ over $\mathfrak{F}$ is a subspace of $\mathscr{V}$ if and only if

$$(X - Y) \in \mathscr{S} \qquad \text{and} \qquad sX \in \mathscr{S} \tag{8}$$

for every element X and Y of $\mathscr{S}$ and every $s \in \mathfrak{F}$.

From (8) it follows that if $(\mathscr{U}_i)_{i \in \alpha}$ is a nonempty family of subspaces $\mathscr{U}_i$ of a vector space $\mathscr{V}$ then the intersection

$$\bigcap_{i \in \alpha} \mathscr{U}_i$$

is a subspace of $\mathscr{V}$.

For $n > 0$, let $V_1, V_2, \ldots, V_n$ be elements of a vector space $\mathscr{V}$ over a field $\mathfrak{F}$ and let $s_1, s_2, \ldots, s_n$ be elements of $\mathfrak{F}$. Then the vector

$$s_1 V_1 + s_2 V_2 + \cdots + s_n V_n = \sum_{i=1}^{n} s_i V_i \tag{9}$$

is called a *finite linear combination* of n vectors $V_1, V_2, \ldots, V_n$ and the scalars $s_1, s_2, \ldots, s_n$ are called the *coefficients* of linear combination (9).

Let us observe that since vector addition is a binary operation, an infinite linear combination is meaningless. Thus, without ambiguity, we may refer to a finite linear combination simply as a *linear combination*. Moreover, when the magnitude of $n > 0$ is immaterial for our purposes, we may denote $\sum_{i=1}^{n} s_i V_i$ simply as $\sum s_i V_i$.

As expected, by a linear combination of vectors $V_1, V_2, \ldots$ we mean $\sum s_i V_i$ for some finitely many vectors $V_i \in \{V_1, V_2, \ldots\}$.

Let $\mathscr{S}$ be a nonempty subset of a vector space $\mathscr{V}$. Clearly, the set of all linear combinations of the elements of $\mathscr{S}$ forms a subspace of $\mathscr{V}$. This subspace is called the *subspace generated* or *spanned* by $\mathscr{S}$ and is denoted by $[\mathscr{S}]$. It is easy to verify that $[\mathscr{S}]$ is equal to the intersection of all the subspaces $\mathcal{U}$ of $\mathcal{V}$ such that $\mathcal{S} \subset \mathcal{U}$. Thus, $[\mathcal{S}]$ is the smallest (with respect to the set-theoretical inclusion $\subset$) subspace of $\mathcal{V}$ containing $\mathcal{S}$.

Let $\mathcal{S} = \{V_1, V_2, \ldots\}$ be a subset of a vector space $\mathcal{V}$. Then $[\mathcal{S}]$ or $[\{V_1, V_2, \ldots\}]$ is often denoted by $[V_1, V_2, \ldots]$ and is called the subspace of $\mathcal{V}$ *generated* (or *spanned*) *by* vectors $V_1, V_2, \ldots$.

From (8) it follows that a nonempty subset $\mathcal{S}$ of a vector space $\mathcal{V}$ is a subspace of $\mathcal{V}$ if and only if for every element U and V of $\mathcal{S}$ every linear combination $sU + tV$ is an element of $\mathcal{S}$.

A vector V is said to be *linearly dependent* on a set $\mathcal{S}$ of vectors if

$$V \in [\mathcal{S}]. \tag{10}$$

Otherwise, V is said to be *linearly independent* of $\mathcal{S}$.

Again if $\mathcal{S} = \{V_1, V_2, \ldots\}$ and (10) holds, it is customary to say that V is *linearly dependent* on the vectors $V_1, V_2, \ldots$; otherwise, V is said to be *linearly independent* of $V_1, V_2, \ldots$.

Clearly, if V is linearly dependent on the vectors $V_1, V_2, \ldots$ then V is equal to a linear combination of the vectors $V_1, V_2, \ldots$.

DEFINITION 2. *A nomempty subset $\mathcal{S}$ of a vector space is said to be a linearly independent set of vectors if no linear combination of distinct elements of $\mathcal{S}$ with nonzero coefficients is equal to the zero vector. Otherwise, $\mathcal{S}$ is said to be a linearly dependent set of vectors.*

Here again we say that the vectors $V_1, V_2, \ldots$ are linearly independent if no linear combination of distinct $V_1, V_2, \ldots$ with nonzero coefficients is equal to the zero vector.

Thus, the set $\{0\}$ which is a nonempty subset of every vector space is linearly dependent. On the other hand, every one-element subset of a vector space which is not equal to $\{0\}$ is linearly independent. Also, we have

REMARK 1. *If $\mathcal{S}$ is a subset of a vector space such that $0 \in \mathcal{S}$ then $\mathcal{S}$ is a linearly dependent set of vectors.*

In view of Definition 2 a necessary and sufficient condition for a nonempty subset $\mathcal{S}$ of a vector space to be linearly independent is that

$$\sum s_i V_i = 0 \qquad \text{if and only if } s_i = 0 \tag{11}$$

with $V_i \in \mathcal{S}$ and $V_i \neq V_j$ for $i \neq j$.

LEMMA 1. *Let $\mathcal{S}$ be a subset of a vector space such that $\mathcal{S}$ has more than one element. Then $\mathcal{S}$ is a linearly independent set of vectors if and* *only if no element of $\mathscr{S}$ is linearly dependent on the remaining elements of $\mathscr{S}$.*

Proof. Let $\mathscr{S}$ be linearly independent. Assume the contrary, that $V_j = \sum_{i \neq j} s_i V_i$ with $V_i \neq V_j$ for $i \neq j$. Then $\sum s_i V_i = 0$ with $s_i = -1 \neq 0$ for $i = j$ which in view of (11) contradicts the linear independence of $\mathscr{S}$.

Next, suppose no element of $\mathscr{S}$ is linearly dependent on the remaining elements of $\mathscr{S}$. Assume the contrary, that $\mathscr{S}$ is linearly dependent and for some $s_k \neq 0$ let $\sum s_i V_i = 0$ with $V_i \neq V_j$ for $i \neq j$. Then, $V_k = \sum_{i \neq k} s_k^{-1} s_i V_i$ contradicting our supposition.

LEMMA 2. *Let $\mathscr{S}$ be a nonempty subset of a vector space. Then $\mathscr{S}$ is linearly independent if and only if every element of $[\mathscr{S}]$ has a unique representation as a linear combination of distinct elements of $\mathscr{S}$.*

Proof. Let $\mathscr{S}$ be linearly independent. Assume the contrary, that an element of $[\mathscr{S}]$ has two distinct representations as a linear combination of distinct elements of $\mathscr{S}$. Then the difference of these two representations, *i.e.*, the zero vector is a linear combination of distinct elements of $\mathscr{S}$ with nonzero coefficients contradicting (11).

Next, suppose every element of $[\mathscr{S}]$ has a unique representation as a linear combination of distinct elements of $\mathscr{S}$. Then since $0 \in [\mathscr{S}]$ and $0 = \sum s_i V_i$ if and only if $s_i = 0$ it follows from (11) that $\mathscr{S}$ is linearly independent.

DEFINITION 3. *A linearly independent set $\mathscr{B}$ of vectors is said to be a basis of a vector space $\mathscr{V}$ if $[\mathscr{B}] = \mathscr{V}$.*

As expected, the notion of basis is also applicable to a subspace $\mathscr{U}$ of a vector space $\mathscr{V}$. Thus, a linearly independent set $\mathscr{W}$ of vectors is a basis of a subset $\mathscr{U}$ of $\mathscr{V}$ if $[\mathscr{W}] = \mathscr{U}$.

Let us also mention that if a subset $\mathscr{G}$ of a vector space $\mathscr{V}$ is such that $[\mathscr{G}] = \mathscr{V}$ then $\mathscr{G}$ is called a set of *generators* of $\mathscr{V}$.

Every vector space $\mathscr{V}$ has a set of generators since $[\mathscr{V}] = \mathscr{V}$. Obviously, a set of generators $\mathscr{G}$ of a vector space $\mathscr{V}$ is a basis of $\mathscr{V}$ if and only if $\mathscr{G}$ is a linearly independent set.

THEOREM 1. *Let $\mathscr{V}$ be a vector space with more than one element and $\mathscr{G}$ be a set of generators of $\mathscr{V}$. Then there exists a basis $\mathscr{B}$ of $\mathscr{V}$ such that $\mathscr{B} \subset \mathscr{G}$.*

Proof. Since $\mathscr{V}$ has more than one element and since $[\mathscr{G}] = \mathscr{V}$ there exists $G \in \mathscr{G}$ such that $G \neq 0$. Thus, $\mathscr{G}$ has a linearly independent subset, namely, $\{G\}$. Consider the set Γ of all linearly independent subsets of $\mathscr{G}$ and partial order Γ by $\subset$. Since Γ is nonempty and since every simply ordered subset Θ of Γ has an upper bound (*e.g.* $\cup \Theta$) by Zorn's lemma Γ has a maximal element $\mathscr{B}$. We claim that $[\mathscr{B}] = \mathscr{V}$. Assume the contrary and let $V \in (\mathscr{V} - [\mathscr{B}])$. Thus, $V \in ([\mathscr{G}] - [\mathscr{B}])$ and $V = \sum\limits_{i \in \alpha} g_i G_i$ with $G_i \in \mathscr{G}$ and with $G_j \notin \mathscr{B}$ for some $j \in \alpha$. But then none of the elements of $\mathscr{B} \cup \{G_j\}$ is linearly dependent on the remaining elements of $\mathscr{B} \cup \{G_j\}$. Hence from Lemma 1, it follows that $\mathscr{B} \cup \{G_j\}$ is a linearly independent subset of $\mathscr{G}$ contradicting the maximality of $\mathscr{B}$. Thus, indeed $[\mathscr{B}] = \mathscr{V}$ and the Theorem is proved.

COROLLARY 1. *Every vector space with more than one element has a basis.*

The proof of the Corollary follows from Theorem 1 since every vector space is its own set of generators.

THEOREM 2. *Let $\mathscr{S}$ be a linearly independent subset of a vector space $\mathscr{V}$ and $\mathscr{B}$ a basis of $\mathscr{V}$. Then there exists a subset $\mathscr{B}_1$ of $\mathscr{B}$ such that $\mathscr{S} \cup \mathscr{B}_1$ is a basis of $\mathscr{V}$.*

Proof. Let Γ be the set of all linearly independent subsets $\mathscr{U}$ of $\mathscr{V}$ such that $\mathscr{S} \subset \mathscr{U}$ and $\mathscr{U} \subset (\mathscr{S} \cup \mathscr{B})$. Then an argument similar to that given in the proof of Theorem 1 shows that a maximal element (with respect to $\subset$) of Γ is equal to $\mathscr{S} \cup \mathscr{B}_1$ for a subset $\mathscr{B}_1$ of $\mathscr{B}$ and that $\mathscr{S} \cup \mathscr{B}_1$ is a basis of $\mathscr{V}$.

LEMMA 3. *Let $V_1, V_2, \ldots, V_n$ be linearly independent vectors. Then*

$$V_2, \ldots, V_n \qquad and \qquad V = \sum s_i V_i \qquad with\ s_1 \neq 0$$

are also linearly independent. Moreover

$$[V_1, V_2, \ldots, V_n] = [V_2, \ldots, V_n, V].$$

Proof. Let

$$0 = a_2 V_2 + \cdots + a_n V_n + t V.$$

Thus,

$$0 = t s_1 V_1 + (a_2 + t s_2) V_2 + \cdots + (a_n + t s_n) V_n.$$

But then (11) implies that

$$t s_1 = (a_2 + t s_2) = \cdots = (a_n + t s_n) = 0$$

and since $s_1 \neq 0$ it follows that $t = 0$ and consequently $0 = a_2 V_2 + \cdots +$

$a_n V_n$ which again in view of (11) implies $a_2 = \cdots = a_n = 0$. Thus, $V_2, \ldots,$ V_n, V are linearly independent. The proof of the second part of the Lemma is straightforward.

THEOREM 3. *Let $\mathscr{B}$ and $\mathscr{C}$ be two bases of a vector space $\mathscr{V}$. Then $\mathscr{B}$ and $\mathscr{C}$ are equipollent, i.e.,*

$$\overline{\overline{\mathscr{B}}} = \overline{\overline{\mathscr{C}}}$$

Proof. Let $B \in \mathscr{B}$. Thus, $B \in \mathscr{V}$ and B is a linear combination of certain elements $C_1, C_2, \ldots, C_n$ of $\mathscr{C}$. Let $B = s_1 C_1 + s_2 C_2 + \cdots + s_n C_n$ with $s_i \neq 0$. In view of Lemma 3,

$$\mathscr{C}_1 = \mathscr{C} \cup \{B\} - \{C_1\}$$

is again a basis of $\mathscr{V}$. Moreover $\overline{\overline{\mathscr{C}}} = \overline{\overline{\mathscr{C}}}_1$.

Thus, $\mathscr{B}$ has a subset $\mathscr{U}_1$ (namely $\{B\}$) such that there exists a basis $\mathscr{C}_1$ of $\mathscr{V}$ with $\overline{\overline{\mathscr{C}}} = \overline{\overline{\mathscr{C}}}_1$ and such that $\mathscr{U}_1$ is a subset of $\mathscr{C}_1$.

Consider the set Γ of all subsets $\mathscr{U}_i$ of $\mathscr{B}$ such that there exists a basis $\mathscr{C}_i$ of $\mathscr{V}$ with $\overline{\overline{\mathscr{C}}} = \overline{\overline{\mathscr{C}}}_i$ and such that $\mathscr{U}_i$ is a subset of $\mathscr{C}_i$.

Clearly, Γ is nonempty. In view of Zorn's lemma it can be easily proved that Γ has a maximal element $\mathscr{U}_m$ (with respect to $\subset$) such that there exists a basis $\mathscr{C}_m$ of $\mathscr{V}$ with $\overline{\overline{\mathscr{C}}} = \overline{\overline{\mathscr{C}}}_m$ and such that $\mathscr{U}_m$ is a subset of $\mathscr{C}_m$.

We claim that $\mathscr{U}_m = \mathscr{B}$. Assume the contrary and let $B_m \in (\mathscr{B} - \mathscr{U}_m)$. Since $B_m \in \mathscr{V}$ and since $\mathscr{C}_m$ is a basis of $\mathscr{V}$ it follows that B_m is a linear combination of certain elements $K_1, K_2, \ldots, K_n$ of $\mathscr{C}_m$ such that, say, K_1 is an element of $\mathscr{C}_m - \mathscr{U}_m$. But then in view of Lemma 3,

$$\mathscr{C}_{m+1} = \mathscr{C}_m \cup \{B_m\} - \{K_1\}$$

is again a basis of $\mathscr{V}$ with $\overline{\overline{\mathscr{C}}} = \overline{\overline{\mathscr{C}}}_{m+1}$. Consequently, $\mathscr{U}_m \cup \{B_m\}$ is an element of Γ contradicting the maximality of $\mathscr{U}_m$. Thus, $\mathscr{U}_m = \mathscr{B}$. Therefore $\mathscr{B}$ is equipollent to a subset of $\mathscr{C}_m$. However, $\overline{\overline{\mathscr{C}}} = \overline{\overline{\mathscr{C}}}_m$ and hence $\mathscr{B}$ is of power less than or equal to $\mathscr{C}$.

Repeating the above argument while interchanging the roles of $\mathscr{B}$ and $\mathscr{C}$ we conclude that $\mathscr{C}$ is also of power less than or equal to $\mathscr{B}$. Consequently, by Cantor-Bernstein theorem $\overline{\overline{\mathscr{B}}} = \overline{\overline{\mathscr{C}}}$, as desired.

Let us observe that there are two algebraic operations involved in a vector space. They are addition of two vectors and multiplication of a scalar by a vector. If $\mathscr{V}$ and $\mathscr{U}$ are two vector spaces over the same field $\mathscr{F}$ then it is natural to call a mapping φ from $\mathscr{V}$ onto $\mathscr{U}$ a *homomorphism from $\mathscr{V}$ onto $\mathscr{U}$* if

$$\varphi(A + B) = \varphi(A) + \varphi(B) \tag{12}$$

$$\varphi(sA) = s\varphi(A) \tag{13}$$

for every element A and B of $\mathscr{V}$ and every element s of $\mathfrak{F}$.

We observe that conditions (12) and (13) can be combined into a single condition, namely,

$$\varphi(sA + tB) = s\varphi(A) + t\varphi(B) \tag{14}$$

for every element A and B of $\mathscr{V}$ and every element s and t of $\mathfrak{F}$.

If φ is a homomorphism from a vector space $\mathscr{V}$ onto a vector space $\mathscr{U}$ then $\mathscr{U}$ is called a *homomorphic image* of $\mathscr{V}$.

A homomorphism φ from a vector space $\mathscr{V}$ onto a vector space $\mathscr{U}$ is called an *isomorphism between $\mathscr{V}$ and $\mathscr{U}$* (or more precisely, an *isomorphism from $\mathscr{V}$ onto $\mathscr{U}$*) if φ is a one-to-one mapping.

Clearly, isomorphism is an equivalence relation in any set of vector spaces over the same field. Moreover, two isomorphic vector spaces are algebraically indistinguishable.

If there exists an isomorphism between two vector spaces $\mathscr{V}$ and $\mathscr{U}$, we denote this by

$$\mathscr{V} \cong \mathscr{U}$$

and we call each an *isomorphic image* of the other.

Let φ be a homomorphism from a vector space $\mathscr{V}$ onto a vector space $\mathscr{U}$. It can be readily verified that the *kernel* (*i.e.*, the set of all vectors X of $\mathscr{V}$ such that $\varphi(X)$ is the zero vector of $\mathscr{U}$) $\mathscr{K}$ of φ is a subspace of $\mathscr{V}$. Moreover,

$$\mathscr{V}/\mathscr{K} \cong \mathscr{U}$$

where in $\mathscr{V}/\mathscr{K}$ addition and multiplication by a scalar of elements of $\mathscr{V}/\mathscr{K}$ are defined in an obvious way.

Conversely, every subspace $\mathscr{K}$ of a vector space $\mathscr{V}$ over a field $\mathfrak{F}$ gives rise (in an obvious way) to the *quotient vector space $\mathscr{V}/\mathscr{K}$* over $\mathfrak{F}$.

Let $\mathscr{V}$ and $\mathscr{W}$ be two vector spaces over the same field $\mathfrak{F}$ and φ be a mapping from $\mathscr{V}$ into $\mathscr{W}$ satisfying (14). Since φ preserves linear combinations it is customary to call φ a *linear mapping* from $\mathscr{V}$ into $\mathscr{W}$. In this connection φ is also called an *into homomorphism* and if φ is one-to-one it is called an *into isomorphism*.

A linear mapping from a vector space $\mathscr{V}$ into $\mathscr{V}$ is usually called a *linear transformation* on $\mathscr{V}$.

Let φ be a linear mapping (*i.e.*, an into homomorphism) from a vector space $\mathscr{V}$ into a vector space $\mathscr{W}$. Then φ need not map a linearly independent subset of $\mathscr{V}$ onto a linearly independent subset of $\mathscr{W}$. However, if φ is one-to-one (*i.e.*, an into isomorphism) then φ maps every linearly independent subset of $\mathscr{V}$ onto a linearly independent subset of $\mathscr{W}$.

Let $\mathscr{V}$ and $\mathscr{W}$ be vector spaces over the same field $\mathfrak{F}$ and $\mathscr{B}$ be a basis of $\mathscr{V}$. If φ is a mapping from $\mathscr{B}$ into $\mathscr{W}$ then in view of (14) there exists a unique linear mapping ψ from $\mathscr{V}$ into $\mathscr{W}$ such that $\varphi = \psi$ on $\mathscr{B}$. Moreover, if φ is a one-to-one mapping from $\mathscr{B}$ into a linearly independent subset of $\mathscr{W}$ then there exists a unique one-to-one linear mapping ψ from $\mathscr{V}$ into $\mathscr{W}$ such that $\varphi = \psi$ on $\mathscr{B}$.

Clearly, if φ is an isomorphism between a vector space $\mathscr{V}$ and a vector space $\mathscr{U}$ then φ maps every basis of $\mathscr{V}$ onto a basis of $\mathscr{U}$.

In the case of vector spaces, the notion parallel to that of the direct sum of rings (*see* page 15) is the supplementary sum of vector spaces.

Consider the vector spaces $\mathscr{U}_1, \mathscr{U}_2, \ldots, \mathscr{U}_n$ over the same field $\mathfrak{F}$ and let $\mathscr{U}$ be the set of all ordered n-tuples $(U_1, U_2, \ldots, U_n)$ with $U_i \in \mathscr{U}_i$. Then the *supplementary sum* (or, more appropriately, the *external supplementary sum*)

$$\mathscr{U}_1 \dotplus \mathscr{U}_2 \dotplus \cdots \dotplus \mathscr{U}_n = \overset{n}{\underset{i=1}{\dotplus}} \mathscr{U}_i \tag{15}$$

of the vector spaces $\mathscr{U}_1, \mathscr{U}_2, \ldots, \mathscr{U}_n$ is defined to be the set $\mathscr{U}$ where addition among the elements of $\mathscr{U}$ is performed coordinatewise and multiplication of an element (*i.e.*, scalar) s $\mathfrak{F}$ by an element $(U_1, U_2, \ldots, U_n)$ of $\mathscr{U}$ is performed according to

$$s(U_1, U_2, \ldots, U_n) = (sU_1, sU_2, \ldots, sU_n). \tag{16}$$

It is readily seen that (15) with addition and multiplication by scalars as defined in the above is a vector space over $\mathfrak{F}$. As expected, in (15) each $\mathscr{U}_i$ is called a *supplementary component* or a *supplementary summand* of $\dotplus \mathscr{U}_i$.

A basis of (15) is easily seen to be the set of all n-tuples $(0_1, 0_2, \ldots, B_i, \ldots, 0_n)$ where B_i is an element of a fixed basis of $\mathscr{U}_i$ and where 0_i is the zero vector of $\mathscr{U}_i$.

We see that every element of $\dotplus \mathscr{U}_i$ is uniquely expressed in terms of elements of each $\mathscr{U}_i$.

Motivated by the above we say that a vector space $\mathscr{V}$ is a *supplementary sum* (or more appropriately, an *internal supplementary sum*) of its subspaces $\mathscr{V}_1, \mathscr{V}_2, \ldots, \mathscr{V}_n$ and we denote this by

$$\mathscr{V} = \mathscr{V}_1 \dotplus \mathscr{V}_2 \dotplus \cdots \dotplus \mathscr{V}_n = \overset{n}{\underset{i=1}{\dotplus}} \mathscr{V}_i \tag{17}$$

if every element of $\mathscr{V}$ has a unique representation as a sum of elements each belonging to a distinct $\mathscr{V}_i$.

Let $\mathscr{S}_1, \mathscr{S}_2, \ldots, \mathscr{S}_n$ be subsets of a vector space $\mathscr{V}$ then as on page 16 we define their sum denoted by

$$\mathscr{S}_1 + \mathscr{S}_2 + \cdots + \mathscr{S}_n = \sum_{i=1}^{n} \mathscr{S}_i$$

as the set of all sums $S_1 + S_2 + \cdots + S_n = \sum_{i=1}^{n} S_i$ with $S_i \in \mathscr{S}_i$.

THEOREM 4. *Let $\mathscr{V}_1, \mathscr{V}_2, \ldots, \mathscr{V}_n$ be subspaces of a vector space $\mathscr{V}$ then* (18) *through* (20) *are equivalent statements.*

$$\mathscr{V} = \overset{n}{\underset{i=1}{\dot{+}}} \mathscr{V}_i. \tag{18}$$

$$\mathscr{V} = \sum_{i=1}^{n} \mathscr{V}_i \quad and \quad \sum_{i=1}^{n} V_i = 0 \quad with \quad V_i \in \mathscr{V}_i \tag{19}$$

implies $V_i = 0$ for $i = 1, 2, \ldots, n$.

$$\mathscr{V} = \sum_{i=1}^{n} \mathscr{V}_i \quad and \quad \mathscr{V}_i \cap \sum_{j \neq i} \mathscr{V}_j = \{0\} \quad for \quad i = 1, 2, \ldots, n. \tag{20}$$

Proof. Let (18) hold. Then clearly, $\mathscr{V} = \sum \mathscr{V}_i$ and $\sum V_i = 0$ implies $V_i = 0$ since, otherwise, 0 will have two distinct representations. Thus, (19) is established.

Next, let (19) hold and let a nonzero element U_i be such that $U_i \in \mathscr{V}_i$ and $U_i = \sum_{j \neq i} V_j$ with $V_j \in \mathscr{V}_j$. Thus,

$$\sum V_k = 0 \text{ with } V_k = -U_i \neq 0 \text{ and } V_k \in \mathscr{V}_k \text{ for } k = 1, 2, \ldots, n.$$

Thus, (20) is established.

Finally, let (20) hold. Let $V \in \mathscr{V}$ be such that $V = V_1 + V_2 + \cdots + V_n = U_1 + U_2 + \cdots + U_n$ with, say, $V_1 \neq U_1$ and $V_i \in \mathscr{V}_i$ and $U_i \in \mathscr{V}_i$. Then $V_1 - U_1 \neq 0$ and $\{(V_1 - U_1)\} \subset (\mathscr{V}_1 \cap \sum_{j \neq 1} \mathscr{V}_j)$ contradicting (20). Thus, (18) is established.

Hence, Theorem 4 is proved.

THEOREM 5. *Let $\mathscr{U}$ be a subspace of a vector space $\mathscr{V}$. Then $\mathscr{U}$ is a supplementary summand of $\mathscr{V}$, i.e.,*

$$\mathscr{V} = \mathscr{U} \dot{+} \mathscr{W}$$

for some subspace $\mathscr{W}$ of $\mathscr{V}$. Moreover,

$$\mathscr{V}/\mathscr{U} \cong \mathscr{W}.$$

Proof. Let $\mathscr{B}$ be a basis of $\mathscr{V}$ and $\mathscr{S}$ a basis of $\mathscr{U}$. In view of Theorem 2, there exists a subset $\mathscr{B}_1$ of $\mathscr{B}$ such that $\mathscr{S} \cup \mathscr{B}_1$ is a basis of $\mathscr{V}$. Let

$\mathcal{W} = [\mathscr{B}_1]$. Then clearly, $\mathcal{V} = \mathcal{U} \dotplus \mathcal{W}$, as desired. The proof of the fact that $\mathcal{V}/\mathcal{U} \cong \mathcal{W}$ is straightforward.

Exercises

1. Prove that in the definition of a vector space given on page 39, condition (4) is independent of conditions (1), (2) and (3).

2. Consider the set of all real numbers $\mathscr{R}$ as a vector space over the field of rational numbers. Prove that every basis of $\mathscr{R}$ is nondenumerable.

3. Prove that any left (or right) unitary module (*see* page 40) over a division ring has a basis (*see* Definition 3).

4. Prove that conditions (12) and (13) together are equivalent to condition (14).

5. Let $\mathscr{P}$ and $\mathscr{Q}$ be submodules of a left module over a ring (*see* page 40). Based on the notations introduced on pages 46 and 48 prove that $(\mathscr{P}+\mathscr{Q})/\mathscr{Q}$ is isomorphic to $\mathscr{P}/(\mathscr{P} \cap \mathscr{Q})$.

6. A (left or right) module $\mathcal{M}$ over a ring (*see* page 40) is said to satisfy the Descending Chain Condition if for every nonincreasing sequence $\mathscr{S}_1 \supset \mathscr{S}_2 \supset \cdots$ of submodules $\mathscr{S}_i$ of $\mathcal{M}$ there exists a natural number n such that $\mathscr{S}_n = \mathscr{S}_{n+1} = \cdots$.

Prove that $\mathcal{M}$ satisfies the Descending Chain Condition if and only if in any nonempty set of submodules of $\mathcal{M}$ there exists a minimal (with respect to $\subset$) submodule.

7. A (left or right) module $\mathcal{M}$ over a ring (*see* page 40) is said to satisfy the Ascending Chain Condition if for every nondecreasing sequence $\mathscr{S}_1 \subset \mathscr{S}_2 \subset \cdots$ of submodules $\mathscr{S}_i$ of $\mathcal{M}$ there exists a natural number n such that $\mathscr{S}_n = \mathscr{S}_{n+1} = \cdots$.

Prove that $\mathcal{M}$ satisfies the Ascending Chain Condition if and only if in every nonempty set of submodules of $\mathcal{M}$ there exists a maximal (with respect to $\subset$) submodule.

8. A submodule $\mathscr{S}$ of a left module over a ring $\mathscr{R}$ (*see* page 40) is called *finitely generated* if there exists a subset $\mathscr{G}$ of $\mathscr{S}$ such that $\mathscr{S}$ is equal to the set of all elements of the form

$$n_1 G_1 + n_2 G_2 + \cdots + n_t G_t + r_1 G_1 + r_2 G_2 + \cdots + r_t G_t$$

where $n_1, n_2, \ldots, n_t$ are natural numbers $r_1, r_2, \ldots, r_t$ elements of $\mathscr{R}$ and $G_1, G_2, \ldots, G_t$ elements of $\mathscr{G}$.

Prove that a left module $\mathcal{M}$ over a ring satisfies the Ascending Chain Condition (*see* Problem 3) if and only if every submodule of $\mathcal{M}$ is finitely generated.

9. Define the notions of the Descending and Ascending Chain Conditions for a ring (by replacing the notions of a submodule by that of a left ideal in the corresponding definitions for a left module given in Problems 2 and 3).

Prove that if a ring $\mathscr{R}$ satisfies the Ascending (Descending) Chain Condition for left ideals then any finitely generated unitary left module (*see* page 40) over $\mathscr{R}$ satisfies the Ascending (Descending) Chain Condition.

10. Let $\mathscr{V}$ be a vector space over a field $\mathfrak{F}$ and $\mathscr{T}$ the set of all linear transformations on $\mathscr{V}$ (*see* page 46). For every element T_1 and T_2 of $\mathscr{T}$ define their sum $T_1 + T_2$ by $(T_1 + T_2)X = T_1(X) + T_2(X)$ for every element X of $\mathscr{V}$. Moreover, for every element s of $\mathfrak{F}$ and every element T of $\mathscr{T}$ define sT by $sT(X) = s(T(X))$ for every element X of $\mathscr{V}$. Prove that $\mathscr{T}$ is a vector space over $\mathfrak{F}$.

11. Consider an abelian group as a left module over the ring of integers (*see* page 40). Prove that every finitely generated abelian group (in the sense of Problem 10) is a supplementary sum (*see* page 47) of a finite number of cyclic subgroups.

12. Let $\mathscr{M}$ be a finitely generated left module (in the sense of Problem 10) over a Euclidean ring (*see* page 27). Prove that $\mathscr{M}$ is a supplementary sum (*see* page 47) of a finite number of submodules each generated by one element.

13. A (left or right) module $\mathscr{M}$ over a ring (*see* page 40) is said to be *irreducible* if $\{0\}$ and $\mathscr{M}$ are the only submodules of $\mathscr{M}$.

Let $\mathscr{M}$ be an irreducible left module over a ring $\mathscr{R}$. Prove that there exists an element M of $\mathscr{M}$ such that every element N of $\mathscr{M}$ is equal to rM for some $r \in \mathscr{R}$, or, that $rM = 0$ for every $M \in \mathscr{M}$ and $r \in \mathscr{R}$.

14. Define a linear transformation on a left module $\mathscr{M}$ over a ring $\mathscr{R}$ imitating the definition of a linear transformation on a vector space given on page 46.

Prove that if $\mathscr{M}$ is an irreducible left module over a ring (*see* Problem 13) and if $rM \neq 0$ for some $r \in R$ and $M \in \mathscr{M}$ then the set of all linear transformations (in the sense of Problem 10) is a division ring.

15. Let $\mathscr{F}$ be a field. Consider $\mathscr{F}$ as a vector space over itself. Characterize the set of all linear transformations (in the sense of Problem 10) on $\mathscr{F}$.

2.2. Finite Dimensional Vector Spaces

In the previous section we proved (Corollary 1) that every vector space with more than one element has a basis and that (Theorem 3) every two bases of a vector space are equipollent (*i.e.*, of the same power). Accordingly, if a vector space $\mathscr{V}$ has a *finite basis* (*i.e.*, a basis with finitely many elements) which has n elements ($n = 1, 2, \ldots$) then every basis of $\mathscr{V}$ also has n elements. Thus, we are justified in introducing the following definition.

DEFINITION 4. *Let $\mathscr{V}$ be a vector space with a basis which has a finite number of elements. Then the number of elements in a basis of $\mathscr{V}$ is called the dimension of $\mathscr{V}$ and $\mathscr{V}$ is called a finite dimensional vector space.*

If a vector space is not finite dimensional then it is called an *infinite dimensional* vector space.

If $\mathscr{V}$ is a vector space then it is customary to write

$$\dim \mathscr{V}$$

instead of *dimension of* $\mathscr{V}$.

It follows from Definitions 2 and 3 that a one-element vector space $\{0\}$ has no basis. Thus

$$\dim \{0\} = 0 \tag{21}$$

and $\{0\}$ is called a *zero dimensional* vector space.

THEOREM 6. *Let* $\mathscr{U}$ *be a subspace of a finite dimensional vector space* $\mathscr{V}$. *Then* $\mathscr{U} \neq \mathscr{V}$ *if and only if*

$$\dim \mathscr{U} < \dim \mathscr{V}. \tag{22}$$

Proof. If $\mathscr{U} = \{0\}$ then the proof follows trivially from (21). Thus, in what follows we assume $\mathscr{U} \neq \{0\}$.

Now, let $\mathscr{U} \neq \mathscr{V}$. In view of Corollary 1, let $\mathscr{S}$ be a basis of $\mathscr{U}$ and $\mathscr{B}$ be a basis of $\mathscr{V}$. By Theorem 2, there exists a subset $\mathscr{B}_1$ of $\mathscr{B}$ such that $\mathscr{S} \cup \mathscr{B}_1$ is a basis of $\mathscr{V}$. Clearly, $\mathscr{B}_1 \neq \emptyset$ since otherwise $\mathscr{U} = [\mathscr{S}] = \mathscr{V}$ contradicting $\mathscr{U} \neq \mathscr{U}$. Let $\dim \mathscr{U} = m$ and $\dim \mathscr{V} = n$. Thus, $\mathscr{S}$ has m elements. By Theorem 3, however, $\mathscr{S} \cup \mathscr{B}_i$ has n elements and since $\mathscr{B}_1 \neq \emptyset$ we see that $m < n$.

Next, let (22) hold. But then $\mathscr{U} \neq \mathscr{V}$ since otherwise $\mathscr{U} = \mathscr{V}$ and by Theorem 3, $\dim \mathscr{U} = \dim \mathscr{V}$ contradicting (22).

In view of Theorem 6 we have

COROLLARY 2. *If* $\mathscr{U}$ *is a subspace of a finite dimensional vector space* $\mathscr{V}$ *then*

$$\dim \mathscr{U} \leq \dim \mathscr{V} \tag{23}$$

LEMMA 4. *Let* $\mathscr{V}$ *be an* $n > 0$ *dimensional vector space. Then every linearly independent subset* $\mathscr{S}$ *of* $\mathscr{V}$ *with n elements is a basis of* $\mathscr{V}$.

Proof. Clearly, $[\mathscr{S}] \subset \mathscr{V}$. On the other hand if $[\mathscr{S}] \neq \mathscr{V}$ then by Theorem 6 we have $n = \dim \mathscr{S} < \dim \mathscr{V} = n$ which is a contradiction. Thus, $[\mathscr{S}] = \mathscr{V}$ and $\mathscr{S}$ is a basis of $\mathscr{V}$.

LEMMA 5. *Let* $\mathscr{V}$ *be an n dimensional vector space. Then every subset* $\mathscr{S}$ *of* $\mathscr{V}$ *with* $m > n$ *elements is linearly dependent.*

Proof. Clearly, $[\mathscr{S}] \subset \mathscr{V}$. If $\mathscr{S}$ is linearly independent then by Corollary 2 we have $m = \dim \mathscr{S} \leq \dim \mathscr{V} = n$ which contradicts $m > n$. Thus, $\mathscr{S}$ is linearly independent, as desired.

51

Using notation introduced on page 47, we have

THEOREM 7. *Let $\mathcal{U}$ and $\mathcal{W}$ be finite dimensional subspaces of a vector space. Then*

$$\dim \mathcal{U} + \dim \mathcal{W} = \dim (\mathcal{U} + \mathcal{W}) + \dim (\mathcal{U} \cap \mathcal{W}). \tag{24}$$

Proof. Without loss of generality we may assume that $\dim (\mathcal{U} \cap \mathcal{W}) > 0$ and that

$$\{V_1, \ldots, V_k\} \tag{25}$$

is a basis of $\mathcal{U} \cap \mathcal{W}$. Since $\mathcal{U} \cap \mathcal{W}$ is a subspace of both $\mathcal{U}$ and $\mathcal{W}$, by Theorem 2, there exists a basis of $\mathcal{U}$ of the form

$$\{V_1, \ldots, V_k, U_1, \ldots, U_m\} \tag{26}$$

and a basis of $\mathcal{W}$ of the form

$$\{V_1, \ldots, V_k, W_1, \ldots, W_n\}. \tag{27}$$

Clearly,

$$\{V_1, \ldots, V_k, U_1, \ldots, U_m, W_1, \ldots, W_n\} \tag{28}$$

is a set of generators of $\mathcal{U} + \mathcal{W}$. We prove that (28) is a linearly independent set and thus a basis of $\mathcal{U} + \mathcal{W}$. Let

$$\sum_{i=1}^{k} s_i V_i + \sum_{j=1}^{m} t_j U_j + \sum_{r=1}^{n} h_r W_r = 0 \tag{29}$$

for some scalars s_i, t_j and h_r. Hence

$$- \sum h_r W_r = \sum s_i V_i + \sum t_j U_j$$

which, in view of (26), implies that

$$\sum h_r W_r \qquad \text{is an element of } \mathcal{U}. \tag{30}$$

On the other hand, in view of (27), we see that

$$\sum h_r W_r \qquad \text{is an element of } \mathcal{W}. \tag{31}$$

Thus, from (30) and (31) it follows that $\sum h_r W_r$ is an element of $\mathcal{U} \cap \mathcal{W}$ and hence by (25), for some scalars c_i, we have

$$\sum h_r W_r = \sum c_i V_i$$

But then, in view of (27), the above equality implies that $h_r = 0$ for $r = 1, \ldots, n$. Consequently, (29) reduces to

$$\sum s_i V_i + \sum t_j U_j = 0$$

which, in view of (26), implies that $s_i = t_j = 0$. Thus,

$$s_i = t_j = h_r = 0 \quad for \quad i = 1, \ldots, k; j = 1, \ldots, m; r = 1, \ldots, n. \quad (32)$$

Clearly, (29) and (32) imply that (28) is a basis of $\mathcal{U} + \mathcal{W}$. But then (24) follows from counting the elements of the sets given in (25), (26), (27) and (28).

COROLLARY 3. *Let $\mathcal{U}$ and $\mathcal{W}$ be subspaces of a finite dimensional vector space $\mathcal{V}$. Then*

$$\mathcal{V} = \mathcal{U} \dotplus \mathcal{W} \quad (33)$$

if and only if $\mathcal{V} = \mathcal{U} + \mathcal{W}$ and

$$\dim \mathcal{V} = \dim \mathcal{U} + \dim \mathcal{W}. \quad (34)$$

Proof. Let (33) hold. Then clearly $\mathcal{V} = \mathcal{U} + \mathcal{W}$ and from (18) and (20) it follows that $\mathcal{U} \cap \mathcal{W} = \{0\}$ which in view of (24) establishes (34).

Next, let $\mathcal{V} = \mathcal{U} + \mathcal{W}$ and (34) hold. Then from (24) it follows that $(\mathcal{U} \cap \mathcal{W}) = \{0\}$ which in view of (20) and (18) establishes (33).

From Corollary 3 we derive by induction

COROLLARY 4. *Let $\mathcal{U}_1, \mathcal{U}_2, \ldots, \mathcal{U}_n$ be subspaces of a finite dimensional vector space $\mathcal{V}$. Then*

$$\mathcal{V} = \mathcal{U}_1 \dotplus \mathcal{U}_2 \dotplus \cdots \dotplus \mathcal{U}_n$$

if and only if $\mathcal{V} = \mathcal{U}_1 + \mathcal{U}_2 + \cdots + \mathcal{U}_n$ and

$$\dim \mathcal{V} = \dim \mathcal{U}_1 + \dim \mathcal{U}_2 + \cdots + \dim \mathcal{U}_n.$$

Based on the notion of isomorphism introduced on page 46 we prove the following theorem.

THEOREM 8. *Let $\mathcal{U}$ and $\mathcal{V}$ be finite dimensional vector spaces over the same field. Then $\mathcal{U}$ and $\mathcal{V}$ are isomorphic if and only if they have the same dimension, i.e.,*

$$\mathcal{U} \cong \mathcal{V} \quad if \ and \ only \ if \quad \dim \mathcal{U} = \dim \mathcal{V}$$

Proof. Let $\mathcal{U} \cong \mathcal{V}$ and φ be an isomorphism from $\mathcal{U}$ onto $\mathcal{V}$. Leaving the trivial case of $\dim \mathcal{U} = 0$ aside, let $\dim \mathcal{U} = n > 1$ and let $\{U_1, U_2, \ldots, U_n\}$ be a basis of $\mathcal{U}$. Since φ preserves linear independence it follows that

$$\varphi(U_i) = V_i \quad for \ i = 1, 2, \ldots, n$$

are n linearly independent vectors of $\mathcal{V}$.

Now, let V be an element of $\mathscr{V}$ such that $\varphi(U) = V$. But then $U = \Sigma\, s_i U_i$ (where s_i is a scalar) and consequently

$$\varphi(U) = \varphi\left(\sum s_i U_i\right) = \sum s_i \varphi(U_i) = \sum s_i V_i.$$

Thus, $\{V_1, V_2, \ldots, V_n\}$ is a basis of $\mathscr{V}$ and $\dim \mathscr{V} = n = \dim \mathscr{U}$.

On the other hand, if $\dim \mathscr{U} = \dim \mathscr{V}$ let ψ by a one-to-one correspondence between a basis $\{U_1, U_2, \ldots, U_n\}$ of $\mathscr{U}$ and a basis $\{V_1, V_2, \ldots, V_n\}$ of $\mathscr{V}$ such that $\psi(U_i) = V_i$. Define $\varphi \Sigma\, s_i U_i = \Sigma\, s_i \psi(U_i) = \Sigma\, s_i V_i$. Clearly, φ is an isomorphism between $\mathscr{U}$ and $\mathscr{V}$ and the theorem is proved.

COROLLARY 5. *Let $\mathscr{U}$ be a subspace of a finite dimensional vector space $\mathscr{V}$. Then*

$$\dim (\mathscr{V}/\mathscr{U}) = \dim \mathscr{V} - \dim \mathscr{U} \tag{35}$$

Proof. By Theorem 5

$$\mathscr{V} = \mathscr{U} \dot{+} \mathscr{W}$$

for some subspace $\mathscr{W}$ of $\mathscr{V}$. But then from Corollary 3 it follows that

$$\dim \mathscr{V} = \dim \mathscr{U} + \dim \mathscr{W} \tag{36}$$

On the other hand, by Theorem 5, $\mathscr{V}/\mathscr{U} \cong \mathscr{W}$ which in view of Theorem 8 and (36) establishes (35), as desired.

Let us observe that if $\mathscr{U}$ and $\mathscr{V}$ are vector spaces over the same field then there exists always a homomorphism from $\mathscr{U}$ into $\mathscr{V}$. For instance, if $\psi(X) = 0$ for every $X \in \mathscr{U}$, then ψ is a homomorphism from $\mathscr{U}$ into $\mathscr{V}$. If φ is a homomorphism from $\mathscr{U}$ into $\mathscr{V}$ then clearly the image of $\mathscr{U}$ under φ, i.e., $\varphi(\mathscr{U})$ is a subspace of $\mathscr{V}$. In this connection we have

LEMMA 6. *Let φ be a homomorphism from a finite dimensional vector space $\mathscr{U}$, over a field $\mathfrak{F}$, into a vector space $\mathscr{V}$ over $\mathfrak{F}$. then*

$$\dim (\mathscr{U}/\mathscr{K}) = \dim \varphi(\mathscr{U}). \tag{36a}$$

Proof. Clearly, φ is a homomorphism from $\mathscr{U}$ onto $\varphi(\mathscr{U})$. Therefore, as mentioned on page 46,

$$\mathscr{U}/\mathscr{K} \cong \varphi(\mathscr{U})$$

which by Theorem 8 yields (36a), as desired.

Thus, Lemma 6 is proved.

Let $\mathscr{B} = \{B_1, B_2, \ldots, B_n\}$ be a basis of an n-dimensional vector space $\mathscr{V}$ over a field $\mathfrak{F}$. Let ψ be a mapping from $\mathscr{B}$ into $\mathscr{V}$ given by

$$
\begin{aligned}
\psi(B_1) &= a_{11}B_1 + a_{12}B_2 + \cdots + a_{1n}B_n \\
\psi(B_2) &= a_{21}B_1 + a_{22}B_2 + \cdots + a_{2n}B_n \\
\cdots &= \cdots \qquad \cdots \qquad \cdots \\
\psi(B_n) &= a_{n1}B_1 + a_{n2}B_2 + \cdots + a_{nn}B_n
\end{aligned} \tag{37}
$$

where $a_{ij} \in \mathfrak{F}$.

It is customary to register the mapping ψ given by (37) in the following matrix form:

$$
\begin{pmatrix} a_{11} & \cdots & a_{1n} \\ a_{21} & \cdots & a_{2n} \\ \cdots & \cdots & \cdots \\ a_{n1} & \cdots & a_{nn} \end{pmatrix} \begin{pmatrix} B_1 \\ B_2 \\ \cdots \\ B_n \end{pmatrix} = \begin{pmatrix} \psi(B_1) \\ \psi(B_2) \\ \cdots \\ \psi(B_n) \end{pmatrix} \tag{38}
$$

As mentioned on page 46, there exists a unique homomorphism φ from $\mathscr{V}$ into $\mathscr{V}$ (also called a *linear transformation* on $\mathscr{V}$ or an *endomorphism* of $\mathscr{V}$) such that $\varphi = \psi$ on $\mathscr{B}$.

Thus, with respect to an (ordered) basis $\{B_1, B_2, \ldots, B_n\}$ of a vector space $\mathscr{V}$ over a field $\mathfrak{F}$ an n by n matrix over $\mathfrak{F}$ uniquely determines a linear transformation on $\mathscr{V}$ as indicated by (38). Conversely, to every linear transformation on $\mathscr{V}$ and an (ordered) basis $\{B_1, B_2, \ldots, B_n\}$ of $\mathscr{V}$ there corresponds a unique n by n matrix over $\mathfrak{F}$, as indicated by (38).

From (38) it follows easily that the matrix (a_{ij}) is nonsingular if and only if the corresponding homomorphism is an isomorphism from $\mathscr{V}$ onto $\mathscr{V}$ (also called an *automorphism* of $\mathscr{V}$).

Also, it can be readily seen that if $\mathscr{K}$ is the kernel of the homomorphism which corresponds to the matrix (a_{ij}) in (38), then

$$
n = \dim \mathscr{V} = \operatorname{rank}(a_{ij}) + \dim \mathscr{K} \tag{39}
$$

where the *rank* of a matrix A is the largest natural number r such that A has an r by r minor (submatrix) M with $\det M \neq 0$.

Exercises

1. Let $\mathscr{U}$ be an n dimensional vector space over a field $\mathfrak{F}$ and $\mathscr{V}$ an $m \geq n$ dimensional vector space over $\mathfrak{F}$. Prove that $\mathscr{U}$ is a homomorphic image of $\mathscr{V}$.

2. Let $\mathscr{U}$ be a subset of an m dimensional vector space $\mathscr{V}$. Prove that $\mathscr{U} = \mathscr{V}$ if and only if $\mathscr{U}$ is an m dimensional subspace of $\mathscr{V}$.

3. Give an example of a finite dimensional vector space $\mathscr{U}$ and a subspace $\mathscr{V}$ and distinct subspaces $\mathscr{M}$ and $\mathscr{N}$ of $\mathscr{U}$ such that

$$
\mathscr{U} = \mathscr{V} \dotplus \mathscr{M} \qquad \text{and} \qquad \mathscr{U} = \mathscr{V} \dotplus \mathscr{N}
$$

4. Based on the notations introduced on pages 42 and 48 prove that

$$
\mathscr{U} + \mathscr{V} = [\mathscr{U} \cup \mathscr{V}]
$$

for every subspace $\mathscr{U}$ and $\mathscr{V}$ of a vector space.

5. Let $\mathscr{V}$ be a finite dimensional vector space and $\mathscr{M}, \mathscr{N}, \mathscr{P}$ and $\mathscr{Q}$ subspaces of $\mathscr{V}$ such that $\mathscr{V} = \mathscr{M} + \mathscr{N} + \mathscr{P} + \mathscr{Q}$. Prove that

$$
\dim \mathscr{V} = \dim \mathscr{M} + \dim \mathscr{N} + \dim \mathscr{P} + \dim \mathscr{Q}
$$

if and only if

$$
(\mathscr{M} \cap \mathscr{N}) = ((\mathscr{M} + \mathscr{N}) \cap \mathscr{P}) = ((\mathscr{M} + \mathscr{N} + \mathscr{P}) \cap \mathscr{Q}) = \{0\}
$$

6. Let $\mathscr{V}_1, \mathscr{V}_2, \ldots, \mathscr{V}_n$ be subspaces of a finite dimensional vector space $\mathscr{V}$ such that

$$\mathscr{V} = \mathscr{V}_1 + \mathscr{V}_2 + \cdots + \mathscr{V}_n.$$

Prove that

$$\dim \mathscr{V} = \dim \mathscr{V}_1 + \dim \mathscr{V}_2 + \cdots + \dim \mathscr{V}_n$$

if and only if

$$\mathscr{V}_i \cap (\mathscr{V}_1 + \cdots + \mathscr{V}_{i-1}, \mathscr{V}_{i+1} + \cdots + \mathscr{V}_n) = \{0\}$$

for $i = 1, 2, \ldots, n$.

7. Let $\mathfrak{F}$ be a finite field with q elements. Determine the number of elements of an m dimensional vector space over $\mathfrak{F}$.

8. Let $\mathscr{V}$ be an m dimensional vector space and $\mathscr{P}$ and $\mathscr{Q}$ be subspaces of $\mathscr{V}$ respectively of dimensions p and q. Prove that if $p + q > m$ then $\mathscr{P}$ and $\mathscr{Q}$ have a nonzero vector in common.

9. Prove (39) on page 55.

10. Let (a_{ij}) be the matrix corresponding to the endomorphism φ and the ordered basis $\{B_1, B_2, \ldots, B_n\}$ of a vector space $\mathscr{V}$. Let $\{V_1, V_2, \ldots, V_n\}$ be a basis of $\mathscr{V}$ such that

$$V_i = \sum b_{ij} B_j.$$

Determine the matrix corresponding to φ and the ordered basis $\{V_1, V_2, \ldots, V_n\}$ of $\mathscr{V}$.

2.3. Matrix Representation of Vectors

Perhaps one of the most useful examples of a vector space is the set $\mathfrak{F}_{mn}$ of all m by n matrices (a_{ij}) over a field $\mathfrak{F}$ with the usual (entrywise) matrix addition and (entrywise) multiplication of a scalar (*i.e.*, an element of $\mathfrak{F}$) by a matrix. These operations are given by:

$$(a_{ij}) + (b_{ij}) = (c_{ij}) \qquad \text{with} \qquad c_{ij} = a_{ij} + b_{ij} \tag{40}$$

and

$$s(a_{ij}) = (sa_{ij}). \tag{41}$$

The fact that $\mathfrak{F}_{mn}$ together with (40) and (41) is a vector space over $\mathfrak{F}$ is easily verified.

Remark 2. *In what follows we shall denote by $E_{ij}(m, n)$ an m by n matrix with 1 on its i-th row and j-th column entry and 0 elsewhere. Moreover, when no confusion is likely to arise $E_{ij}(m, n)$ shall be denoted simply by E_{ij}.*

Next, we observe that the vector space $\mathfrak{F}_{mn}$ is finite dimensional. In fact,

$$\dim \mathfrak{F}_{mn} = mn. \tag{42}$$

This is easily seen since the *mn* matrices E_{ij} of $\tilde{\mathfrak{F}}_{mn}$ form a basis of $\tilde{\mathfrak{F}}_{mn}$. Indeed, every *m* by *n* matrix (a_{ij}) has a unique representation as a linear combination of the matrices E_{ij} given by:

$$(a_{jj}) = \sum_{i,j} a_{ij}E_{ij}. \tag{43}$$

For instance, in the case of 1 by 3 matrices we have

$$(a, b, c) = a(1, 0, 0) + b(0, 1, 0) + c(0, 0, 1)$$

and in the case of 2 by 2 matrices we have

$$\begin{pmatrix} q & r \\ s & t \end{pmatrix} = q \begin{pmatrix} 1 & 0 \\ 0 & 0 \end{pmatrix} + r \begin{pmatrix} 0 & 1 \\ 0 & 0 \end{pmatrix} + s \begin{pmatrix} 0 & 0 \\ 1 & 0 \end{pmatrix} + t \begin{pmatrix} 0 & 0 \\ 0 & 1 \end{pmatrix}$$

LEMMA 7.   *Let* $\mathcal{V}$ *be a* $q > 0$ *dimensional vector space over a field* $\tilde{\mathfrak{F}}$ *and let* $mn \geq q$ *where m and n are natural numbers. Then* $\mathcal{V}$ *is isomorphic to a subspace of the vector space* $\tilde{\mathfrak{F}}_{mn}$ *of all m by n matrices over* $\tilde{\mathfrak{F}}$.

Proof.   Let $\{B_1, B_2, \ldots, B_q\}$ be a basis of $\mathcal{V}$. In view of (42), let $\{M_1, M_2, \ldots, M_q\}$ be any *q*-element subset of $\{E_{11}, E_{12}, \ldots, E_{mn}\}$. Consider the mapping ψ given by

$$\psi(B_i) = M_i, \quad i = 1, 2, \ldots, q \tag{44}$$

Then clearly, in view of (43), the mapping φ given by

$$\varphi\left(\sum s_i B_i\right) = \sum s_i \psi(B_i) = \sum s_i M_i$$

is an isomorphism (*see* page 46) between $\mathcal{V}$ and the subspace of $\tilde{\mathfrak{F}}_{mn}$ spanned by $\{M_1, M_2, \ldots, M_n\}$.

Thus, Lemma 7 is proved.

Accordingly, every 2 dimensional vector space over $\tilde{\mathfrak{F}}$ is isomorphic to a subspace of all 3 by 4 matrices over $\tilde{\mathfrak{F}}$. This can be seen by considering the set of all 3 by 4 matrices, say, of the form

$$\begin{pmatrix} 0 & 0 & 0 & 0 \\ 0 & s & 0 & 0 \\ 0 & 0 & 0 & t \end{pmatrix}$$

where *s* and *t* are elements of $\tilde{\mathfrak{F}}$.

Since $1q \geq q$, by Lemma 7, we have

COROLLARY 6.   *Every* $q > 0$ *dimensional vector space over* $\tilde{\mathfrak{F}}$ *is isomorphic to the set of all 1 by q matrices over* $\tilde{\mathfrak{F}}$.

We observe also that Lemma 7 and Corollary 6 are direct consequences of Theorem 8.

In view of Lemma 7, we may represent a vector space $\mathscr{V}$ by a suitable set of matrices. However, it should be emphasized that there is no unique representation of this kind since any such representation, as (44) shows, depends on a particular choice of a basis of $\mathscr{V}$ and on a particular choice of the corresponding set of linearly independent matrices.

Now, let us consider the set $\mathfrak{F}_n$ of all n by n matrices over a field $\mathfrak{F}$ and let M be an element of $\mathfrak{F}_n$ and I be the n by n unit matrix. Since $\mathfrak{F}_n$ is an n^2 dimensional vector space over $\mathfrak{F}$, by Lemma 5, the $n^2 + 1$ vectors

$$M^{n^2}, \quad M^{n^2-1}, \ldots, M^2, \quad M, \quad I \tag{45}$$

are linearly dependent. Clearly, in (45) M^k is the k-times product of the matrix M by itself [*see* (50), page 29].

Thus, there exist $n^2 + 1$ scalars $s, t, \ldots, a, b, c$ such that

$$sM^{n^2} + tM^{n^2-1} + \cdots + aM^2 + bM + cI = 0_n \tag{46}$$

where 0_n is the n by n zero matrix. In view of (46) it is customary to say that the matrix M satisfies the polynomial

$$sIx^{n^2} + tIx^{n^2-1} + \cdots + aIx^2 + bIx + cI$$

or that the matrix M satisfies the equation

$$sIx^{n^2} + tIx^{n^2-1} + \cdots + aIx^2 + bIx + cI = 0_n.$$

From the above it follows that *every n by n matrix over a field $\mathfrak{F}$ satisfies a polynomial of degree $\leq n^2$ whose coefficients are scalar n by n matrices over $\mathfrak{F}$.*

A natural question arises. For $n > 1$, is there a smaller lower bound than n^2 for the degrees of polynomials whose coefficients are n by n scalar matrices over $\mathfrak{F}$ and which are satisfied by n by n matrices over $\mathfrak{F}$? A positive answer to this question is given by the Cayley-Hamilton theorem below.

First, however, let us observe that an m by n matrix with entries in the polynomial ring $\mathfrak{F}[x]$ can be regarded as a polynomial in the indeterminate x whose coefficients are m by n matrices over $\mathfrak{F}$.

For instance, in the case of the following 2 by 3 matrix we have

$$\begin{pmatrix} ax^2 + bx & cx & d \\ ex+g & hx^2 & 0 \end{pmatrix} = \begin{pmatrix} a & 0 & 0 \\ 0 & h & 0 \end{pmatrix} x^2 + \begin{pmatrix} b & c & 0 \\ e & 0 & 0 \end{pmatrix} x + \begin{pmatrix} 0 & 0 & d \\ g & 0 & 0 \end{pmatrix}.$$

Next, let M be an n by n matrix over a field $\mathfrak{F}$. Consider the n by n matrix

$$Ix - M \tag{47}$$

over the ring $\mathfrak{F}[x]$. By (55) on page 32, we have

$$(Ix - M)(\operatorname{adj}(Ix - M)) = (\det(Ix - M))I \tag{48}$$

where I is the unit n by n matrix.

Let us observe that each factor of the left side of $=$ in (48) as well as the right side of $=$ in (48) is an n by n matrix with entries in the polynomial ring $\mathfrak{F}[x]$. Let each factor on the left side of $=$ in (48) as well as the right side of $=$ in (48) be expressed as a polynomial in the indeterminate x with matrix coefficients. Thus,

$$Ix - M \qquad \text{is expressed as} \qquad Ix - M;$$
$$\text{adj}\,(Ix - M) \qquad \text{is expressed as} \qquad Ix^{n-1} + \cdots + \text{adj}\,(-M)$$
$$(\det\,(Ix - M))I \qquad \text{is expressed as} \qquad Ix^n + \cdots + (\det\,(-M))I.$$

From the above in view of (48) and the properties of matrix addition and multiplication we obtain

$$(Ix - M)(Ix^{n-1} + \cdots + \text{adj}\,(-M)) = Ix^n + \cdots + (\det\,(-M))I. \quad (49)$$

Since in (49) the symbol x is an indeterminate we may regard (49) as an equality between two polynomials over the ring $\mathfrak{F}_n$ of all n by n matrices over $\mathfrak{F}$. However, as (49) shows, $(Ix - M)$ is a left divisor of the polynomial in either side of $=$ in (49). Therefore, by Lemma 7 on page 25 we have

$$M^n + \cdots + (\det\,(\quad M))I = 0_n. \quad (50)$$

It is customary to call $Ix - M$ the *characteristic matrix* of the matrix M and $\det\,(Ix - M)$ the *characteristic determinant* of M. Moreover, if $(\det\,(Ix - M))I$ is regarded as a polynomial over the ring $\mathfrak{F}_n$ of all n by n matrices over $\mathfrak{F}$, *i.e.*,

$$Ix^n + \cdots + (\det\,(-M))I \quad (51)$$

then (51) is called the *characteristic polynomial* of M and

$$Ix^n + \cdots + (\det\,(-M))I = 0_n \quad (52)$$

is called the *characteristic equation* of M.

Comparing (50), (52) and (51) we have

THEOREM 9 (Cayley-Hamilton). *Every n by n matrix over a field $\mathfrak{F}$ (or even over a commutative ring with unity) is a root of its characteristic polynomial which is a monic polynomial of degree n and whose coefficients are n by n scalar matrices over $\mathfrak{F}$.*

In view of Theorem 9, we see that in the n^2 dimensional vector space $\mathfrak{F}_n$ of all n by n matrices over a field $\mathfrak{F}$, for every n by n matrix M the $n + 1$ matrices

$$M^n, \quad M^{n-1}, \ldots, M^2, \quad M, \quad I \quad (53)$$

are linearly dependent.

According to Theorem 9, every n by n matrix over a field $\mathfrak{F}$ satisfies a polynomial of degree n whose coefficients are n by n scalar matrices over $\mathfrak{F}$. However, as shown by Theorem 10 below, for $n > 1$ it is possible for an n by n matrix over $\mathfrak{F}$ to satisfy a polynomial of degree $< n$ whose coefficients are n by n scalar matrices over $\mathfrak{F}$.

Recalling the definition of the invariant factors of a matrix (*see* pages 34 and 36) we have

LEMMA 8.　*Let $\mathfrak{F}$ be a field and M an n by n matrix over $\mathfrak{F}$. Then the characteristic determinant of M is equal to the product of the invariant factors of the characteristic matrix $Ix - M$ of M. Moreover, no invariant factor of $Ix - M$ is equal to zero.*

Proof.　By Lemma 12 on page 36, there exist nonsingular matrices P and Q over $\mathfrak{F}[x]$ such that

$$(Ix - M) = P \begin{pmatrix} f_1(x) & 0 & \cdot & 0 \\ 0 & f_2(x) & \cdot & 0 \\ \cdot & & \cdot & \cdot \\ 0 & 0 & 0 & f_n(x) \end{pmatrix} Q \tag{54}$$

where $f_1(x), f_2(x), \ldots, f_n(x)$ are the invariant factors of $Ix - M$.

From (54) and the fact that $f_i(x)$'s are 0, 1 or monic polynomials in x we have

$$\det(Ix - M) = x^n + \cdots + \det(-M) = f_1(x)f_2(x)\ldots f_n(x) \tag{55}$$

from which the conclusion of Lemma 8 follows.

Recalling the definition of the first nonzero invariant factor of a matrix (*see* page 35) we have

THEOREM 10.　*Let $\mathfrak{F}$ be a field and M an n by n matrix over $\mathfrak{F}$. Let*

$$x^m + \cdots + a_1 x + a_0 \tag{56}$$

be the first invariant factor of the characteristic matrix $(Ix - M)$ of M. Then the polynomial

$$Ix^m + \cdots + a_1 Ix + a_0 I \tag{57}$$

is such that $m \geq 1$ and is the unique monic polynomial of least degree whose coefficients are n by n scalar matrices over $\mathfrak{F}$ and

$$M^m + \cdots + a_1 M + a_0 I = 0_n. \tag{58}$$

Proof.　By Lemma 12 on page 36, each $f_i(x)$ in (55) divides $f_{i+1}(x)$. Moreover, since $n \geq 1$, in view of (55), we see that $m \geq 1$.

On the other hand, from the proof of Lemma 10 on page 33 it follows that the first invariant factor of $Ix - M$, *i.e.*, (56) is equal to

$$(\det (Ix - M)) g^{-1}(x) \tag{59}$$

where $g(x)$ is the g.c.d. (*see* Problem 5 on page 38) of the elements of the matrix adj $(Ix - M)$. Thus, in particular

$$\text{adj } (Ix - M) = g(x)(Ip(x)) \tag{60}$$

where $g(x)$ is the g.c.d. (*see* Problem 5 on page 38) of the elements of the But then from (48), (57), (59) and (60) it follows that

$$(Ix - M)(Ip(x)) = Ix^m + \cdots + a_1 Ix + a_0 I. \tag{61}$$

From (61) as in the case of (49) and (50) we derive

$$M^m + \cdots + a_1 M + a_0 I = 0_n$$

which establishes (58).

Next, let $Ix^k + \cdots + b_1 Ix + b_0 I$ be the unique monic polynomial of least degree whose coefficients are n by n scalar matrices over $\mathfrak{F}$ and such that

$$M^k + \cdots + b_1 M + b_0 I = 0_n. \tag{62}$$

But then in view of (58), (62) and the Euclidean Algorithm (*see* page 24) we derive

$$Ix^m + \cdots + a_1 Ix + a_0 I = (Ix^k + \cdots + b_1 Ix + b_0 I)h(x). \tag{63}$$

Also, from (62) in view of Lemma 7 on page 25 we have

$$Ix^k + \cdots + b_1 Ix + b_0 I = (Ix - M)(It(x)) \tag{64}$$

Comparing (61) with (63) and (64) we obtain

$$(Ix - M)(Ip(x)) = (Ix - M)(h(x)(It(x)))$$

which in view of the Euclidean Algorithm (*see* page 24) implies

$$Ip(x) = h(x)(It(x)). \tag{65}$$

However the entries of $Ip(x)$ are elements of $\mathfrak{F}[x]$ whose g.c.d is 1 and therefore from (65) it follows that $h(x)$ is an element of $\mathfrak{F}$ and since (63) is an equality between two monic polynomials we must have $h(x) = 1$. But then from (63) it follows that (57) is indeed the unique monic polynomial of least degree whose coefficients are n by n scalar matrices over $\mathfrak{F}$ and such that (58) is valid.

COROLLARY 7. *Let M be an n by n matrix with entries in a field $\mathfrak{F}$, and, let the degree of the first invariant factor of M be m. Then $0 < m \leqslant$*

n. Moreover the m matrices

$$I, \quad M, \quad M^2, \ldots, M^{m-1}$$

are linearly independent (as elements of the vector space $\mathfrak{F}_n$ over $\mathfrak{F}$) whereas the $M+1$ matrices

$$I, \quad M, \quad M^2, \ldots, M^{m-1}, \quad M^m$$

are linearly dependent.

In view of Theorem 10, we introduce

DEFINITION 5. *Let M be an n by n matrix with entries in a field and let $f_1(x)$ be the first invariant factor of the matrix $Ix - M$. Then $f_1(x)$ is called the minimum polynomial (or the minimum function) of M and $f_1(x) = 0$ is called the minimum equation of M.*

In view of Theorem 10 it is customary to say that a matrix satisfies its minimum equation. This, of course, is in the sense of (58).

REMARK 3. *Instead of (51), usually det $(Ix - M)$ is called the characteristic polynomial of the matrix M. Likewise, instead of (52), usually det $(Ix - M) = 0$ is called the characteristic equation of the matrix M.*

In view of Theorem 9 it is customary to say that a matrix satisfies its characteristic equation. This, again is in the sense of (50).

Let M be an n by n matrix over a field $\mathfrak{F}$ and let

$$x^n + c_{n-1}x^{n-1} + \cdots + c_1 x + c_0 \qquad \text{and} \qquad x^m + \cdots + a_1 x + a_0$$

be respectively the characteristic equation and the minimum equation of M. Then as indicated by (49) and as easily seen

$$c_0 = \det(-M) \qquad \text{and} \qquad c_{n-1} = -\operatorname{tr} M \qquad (66)$$

where $\operatorname{tr} M$ is the *trace* of the matrix M, *i.e.*, the sum of the diagonal entries of M. Moreover, M (as an element of the total matrix ring $\mathfrak{F}_n$) has the multiplicative inverse M^{-1} if and only if $c_0 \neq 0$ or equivalently $a_0 \neq 0$ in which case in view of (50) and (58) we have

$$M^{-1} = -c_0^{-1}(M^{n-1} + \cdots + c_1 I) = -a_0^{-1}(M^{m-1} + \cdots + a_1 I). \quad (67)$$

Let us give an example. Consider the matrix M (with entries, say, in the field of real numbers) given by

$$M = \begin{pmatrix} 0 & 2 & -1 \\ 1 & -1 & 1 \\ 1 & -2 & 2 \end{pmatrix}.$$

The characteristic matrix of M is

$$Ix - M = \begin{pmatrix} x & -2 & 1 \\ -1 & x+1 & -1 \\ -1 & 2 & x-2 \end{pmatrix}.$$

Using transformations of type (i), (ii) and (iii) described on page 33, we diagonalize $Ix - M$ as follows

$$\begin{pmatrix} x & 2 & -1 \\ 1 & 1 & 0 \\ 1 & 0 & 0 \end{pmatrix}\begin{pmatrix} x & -2 & 1 \\ -1 & x+1 & -1 \\ -1 & 2 & x-2 \end{pmatrix}\begin{pmatrix} 1 & 0 & 0 \\ -1 & 1 & 0 \\ -x-2 & 2 & 1 \end{pmatrix} = \begin{pmatrix} x^2-1 & 0 & 0 \\ 0 & x-1 & 0 \\ 0 & 0 & 1 \end{pmatrix}.$$

Thus, the invariant factors of $Ix - M$ are:

$$x^2 - 1, \quad x - 1, \quad 1$$

and the elementary divisors of $Ix - M$ are:

$$x+1, \quad x-1, \quad x-1, \quad 1.$$

The minimum polynomial of the matrix M is

$$x^2 - 1$$

and the characteristic polynomial of M is:

$$(x^2 - 1)(x - 1) = x^3 - x^2 - x + 1.$$

As expected, $M^2 - I = 0$, i.e., M satisfies its minimum equation $x^2 - 1 = 0$. Clearly, $M^{-1} = M$.

We close this section by observing that the invariant factors of the n by n matrix $Ix - A$ where the n by n matrix A is given as:

$$A = \begin{pmatrix} 0 & 1 & 0 & \cdots & 0 & 0 \\ 0 & 0 & 1 & \cdots & 0 & 0 \\ \cdot & \cdot & \cdot & \cdots & \cdot & \cdot \\ 0 & 0 & 0 & \cdots & 0 & 1 \\ -a_0 & -a_1 & -a_2 & \cdots & -a_{n-2} & -a_{n-1} \end{pmatrix} \qquad (68)$$

are:

$$x^n + a_{n-1}x^{n-1} + a_{n-2}x^{n-2} + \cdots + a_2 x^2 + a_1 x + a_0, 1, 1, \ldots, 1. \qquad (69)$$

This can be easily verified by diagonalizing $Ix - A$ by transformations of type (i), (ii) and (iii) described on page 33.

Clearly, the minimum polynomial of matrix A is the first invariant factor given in (69).

The n by n matrix A given by (68) is called the *companion matrix* of the minimum polynomial of A given in (69).

Exercises

1. Prove that the set of all m by n matrices (a_{ij}) over a field $\mathfrak{F}$ together with (40) and (41) is an mn dimensional vector space over $\mathfrak{F}$. Prove also that the mn matrices $E_{ij(m,n)}$ described on page 56 form a basis of this vector space.

2. Let tr M denote the trace of a square matrix as described on page 62. Prove that for every square matrix A and B over $\mathfrak{F}$ and every element s of $\mathfrak{F}$:

$$\text{tr}\,(sA) = s\,(\text{tr}\,A), \qquad \text{tr}\,(A+B) = \text{tr}\,A + \text{tr}\,B, \qquad \text{tr}\,(AB) = \text{tr}\,(BA).$$

Prove also if A is nonsingular then tr $(ABA^{-1}) = \text{tr}\,B$.

3. Let $x^n + c_{n-1}x^{n-1} + \cdots + c_1 x + c_0$ be the characteristic polynomial of an n by n matrix M. Prove (66) on page 62.

4. Prove that a square matrix M is nonsingular if and only if the constant term a_0 of the minimum polynomial of M is nonzero. Also prove (67) on page 62.

5. Let A be the companion matrix of a monic polynomial $p(x)$ of degree n. Prove that the matrix $Ix - A$ has n nonzero invariant factors such that the first invariant factor is equal to $p(x)$ and each of the remaining $n-1$ invariant factors is equal to 1.

6. Determine the invariant factors of each of the following matrices

$$\begin{pmatrix} x-1 & x \\ x+1 & 1 \end{pmatrix}, \quad \begin{pmatrix} x-2 & 1 \\ 0 & x \end{pmatrix}, \quad \begin{pmatrix} x-3 & -2 \\ x-5 & x-1 \end{pmatrix}.$$

7. Determine the characteristic and minimum polynomials of each of the following matrices

$$\begin{pmatrix} -1 & -1 \\ 2 & 1 \end{pmatrix}, \quad \begin{pmatrix} 2 & -1 \\ 0 & 0 \end{pmatrix}, \quad \begin{pmatrix} 1 & 2 \\ 4 & 1 \end{pmatrix}.$$

8. Determine the invariant factors of the characteristic matrix of each of the following matrices

$$\begin{pmatrix} 1 & 0 \\ 0 & 1 \end{pmatrix}, \quad \begin{pmatrix} 1 & 1 \\ 0 & 1 \end{pmatrix}, \quad \begin{pmatrix} 1 & 1 \\ 1 & 1 \end{pmatrix}.$$

9. Determine a matrix A such that the invariant factors of $Ix - A$ are $x-1$ and $x-1$. Determine a matrix B such that the invariant factors of $Ix - B$ are $(x-1)^2$ and 1. Determine a matrix C such that the invariant factors of $Ix - C$ are $x^2 - 2x$ and 1.

10. Determine the multiplicative inverse of the matrix

according to (67) on page 62. $\qquad \begin{pmatrix} 3 & 2 \\ 4 & 3 \end{pmatrix}$

CHAPTER 3

Linear Associative Algebras

3.1. Linear Associative Algebras

It is customary to call a vector space $\mathscr{A}$ a *linear algebra* when besides addition of vectors and multiplication of vectors by scalars, a third operation, namely, multiplication of vectors by vectors (satisfying certain conditions) is also defined in $\mathscr{A}$. Since a linear algebra is *a priori* a vector space, many of the definitions, terminologies and results which were introduced and obtained in Chapter 2 will be used in connection with a linear algebra. Thus, in particular, by a basis and the dimension of an algebra $\mathscr{A}$ we shall mean a basis and the dimension of $\mathscr{A}$ in the sense of Definition 3 on page 43 and Definition 4 on page 50.

DEFINITION 1. *Let $\mathscr{A}$ be an n dimensional vector space over a field $\mathfrak{F}$. Then $\mathscr{A}$ is called an n dimensional linear associative algebra over $\mathfrak{F}$ if for every element X and Y of $\mathscr{A}$ there is a unique element of $\mathscr{A}$ denoted by XY and called the product of X and Y (in this order) such that*

$$(sX)(tY) = (st)(XY) \tag{1}$$

for every element s and t of $\mathfrak{F}$.

$$Z(X+Y) = ZX + ZY \qquad \text{and} \qquad (X+Y)Z = XZ + YZ \tag{2}$$

$$(XY)Z = X(YZ) \tag{3}$$

for every element Z of $\mathscr{A}$.

Clearly, in the above, say, sX is the product of scalar s by vector X and, say, $X+Y$ is the sum of vectors X and Y. As expected, these operations satisfy the conditions set forth in Definition 1 on page 39.

According to (2) in a finite dimensional linear associative algebra, multiplication is distributive with respect to addition. Moreover, according to (3), multiplication is associative.

If in Definition 1, conditions of finite dimensionality and (3) are dropped, then the resulting system is called a *linear algebra*. Although many of the results which are obtained in the sequel are valid in the case of linear algebras, nevertheless, we shall deal exclusively with finite dimensional linear associative algebras.

Thus *in what follows, unless otherwise specified, we shall refer to a finite dimensional linear associative algebra $\mathscr{A}$ over a field $\mathfrak{F}$ as an algebra over $\mathfrak{F}$. Moreover, if the specific properties of the field $\mathfrak{F}$ are immaterial for our purposes, we shall refer to $\mathscr{A}$ simply as an algebra.*

Since an algebra $\mathscr{A}$ over $\mathfrak{F}$ is *a priori* a vector space over $\mathfrak{F}$, we shall occasionally continue to call an element of $\mathscr{A}$ a vector and an element of $\mathfrak{F}$ a scalar. Moreover, as mentioned on page 39, we shall continue to denote algebras and their subsets by script letters; the underlying fields by German letters; scalars by small letters, and, the elements of algebras (*i.e.*, vectors) by capital letters. Naturally, these items shall be denoted also by suitable configurations using appropriate letters. The symbol 0, as usual, will denote both the zero vector and the zero scalar.

In view of Definition 1 on page 39 and the above definition, we see that an algebra is also a ring. Thus, many of the definitions, terminologies and results which were introduced and obtained in Chapter 1 will be used in connection with an algebra.

Let us observe that conditions (1) and (2) can be combined into

$$Z(sX + tY) = sZX + tZY \qquad \text{and} \qquad (sX + tY)Z = sXZ + tYZ. \qquad (4)$$

In view of (4) it is customary to say that in an algebra multiplication is *bilinear*.

An algebra $\mathscr{A}$ is called *commutative* if $XY = YX$ for every element X and Y of $\mathscr{A}$. It is easily seen that (1), (2) and (3) do not imply the commutativity of $\mathscr{A}$.

Let $\mathscr{A}$ be an algebra over $\mathfrak{F}$ and let $\{A_1, A_2, \ldots, A_n\}$ be a basis of $\mathscr{A}$. Clearly, each $A_i A_j$ is an element of $\mathscr{A}$ and as such each $A_i A_j$ has a unique representation as a linear combination of distinct A_k's (*see* Lemma 2, page 43). Let

$$A_i A_j = \sum_{k=1}^{n} a_{ij}^k A_k \qquad (i, j, k = 1, 2, \ldots, n). \qquad (5)$$

Thus, corresponding to a basis $\{A_1, A_2, \ldots, A_n\}$ of $\mathscr{A}$ there exists a unique set of n^3 scalars (*i.e.*, elements of $\mathfrak{F}$) a_{ij}^k satisfying (5).

The n^3 scalars a_{ij}^k are called the *multiplication constants* or the *structure constants* of $\mathscr{A}$ with respect to the basis $\{A_1, A_2, \ldots, A_n\}$ of $\mathscr{A}$.

Let us observe that multiplication in $\mathscr{A}$ is completely determined in terms of a basis $\{A_1, A_2, \ldots, A_n\}$ of $\mathscr{A}$ and the corresponding n^3 multiplication constants a_{ij}^k. This is because if P and Q are elements of $\mathscr{A}$ such that

$$P = \sum p_i A_i \qquad \text{and} \qquad Q = \sum q_j A_j$$

then by (1), (2) and (5) we have

$$PQ = \sum_{i,j} p_i q_j A_i A_j = \sum_{i,j,k} p_i q_j a_{ij}^k A_k. \tag{6}$$

Let $\{A_1, A_2, \ldots, A_n\}$ be a basis of an algebra $\mathscr{A}$. We call the *multiplication table* of $\mathscr{A}$ with respect to the basis $\{A_1, A_2, \ldots, A_n\}$ a table which (in terms of the corresponding structure constants) gives each of the products $A_i A_j$ as a linear combination of A_k's.

For instance if $\{A_1, A_2\}$ is a basis of an algebra $\mathscr{A}$ then the multiplication table of $\mathscr{A}$ is given by:

	A_1	A_2	
A_1	$a_{11}^1 A_1 + a_{12}^1 A_2$	$a_{11}^2 A_1 + a_{12}^2 A_2$	(7)
A_2	$a_{21}^1 A_1 + a_{22}^1 A_2$	$a_{21}^2 A_1 + a_{22}^2 A_2$	

where a_{ij}^k for $i, j, k = 1, 2$ are the multiplication constants of $\mathscr{A}$ with respect to the basis $\{A_1, A_2\}$.

The notation of the structure constants can be simplified if instead of $\{A_1, A_2\}$ we choose $\{A, B\}$ to denote a basis of $\mathscr{A}$. In this case table (7) acquires the following form:

	A	B	
A	$a_{11}A + a_{12}B$	$b_{11}A + b_{12}B$	(8)
B	$a_{21}A + a_{22}B$	$b_{21}A + b_{22}B$	

Naturally, it is understood that the product PQ of the elements P and Q of $\mathscr{A}$ is obtained from the multiplication table via (6).

Let us observe that every algebra $\mathscr{A}$ has a unique multiplication table with respect to a basis of $\mathscr{A}$. However, it is not necessarily the case that a multiplication table for a basis of a vector space $\mathscr{A}$ determines an algebra in $\mathscr{A}$. This is because an arbitrary multiplication table for a basis of a vector space may fail to be associative. In this connection we have the following lemma.

LEMMA 1. *Let $\{A_1, A_2, \ldots, A_n\}$ be a basis of a vector space $\mathscr{A}$ over $\mathfrak{F}$ and let*

$$A_i A_j = \sum a_{ij}^k A_k \qquad (i, j, k = 1, 2, \ldots, n) \tag{9}$$

where a_{ij}^k are elements of $\mathfrak{F}$.

Then $\mathscr{A}$ is an algebra whose multiplication table with respect to the basis $\{A_1, A_2, \ldots, A_n\}$ is given by (9) if and only if

$$(A_i A_j) A_k = A_i (A_j A_k). \qquad (i, j, k = 1, 2, \ldots, n) \tag{10}$$

Moreover, $\mathscr{A}$ is commutative if and only if

$$A_i A_j = A_j A_i. \qquad (i, j = 1, 2, \ldots, n). \qquad (11)$$

Proof. If $\mathscr{A}$ is an algebra then (10) follows from (3). Moreover, (11) follows from the commutativity of $\mathscr{A}$. Conversely, in view of (9) it can be easily verified that (10) implies (1), (2) and (3). Moreover, (11) obviously implies the commutativity of $\mathscr{A}$.

In what follows we shall often refer to a multiplication table of an algebra, especially, in giving examples of algebras.

For instance, the following is the multiplication table with respect to the basis $\{A, B\}$ of a two dimensional algebra $\mathscr{A}$ over the field of rational numbers

	A	B
A	$(3.4)A + (-2.4)B$	$(-0.6)A + (1.6)B$
B	$(-0.6)A + (1.6)B$	$(0.4)A + (0.6)B$

$$(12)$$

It can be easily verified that the above table is associative.

Clearly, the algebra $\mathscr{A}$ under consideration is commutative since $AB = BA$ as shown in (12). Now, if the elements P and Q of $\mathscr{A}$ are such that

$$P = A + B \qquad \text{and} \qquad Q = A - B$$

then in view of table (12) we have

$$PQ = A^2 - B^2 = 3A - 3B.$$

Since in devising a multiplication table of an algebra the associativity of the table is the only nontrivial restriction, we see immediately that the operational table of any finite group can be considered as a multiplication table of an algebra.

Thus, for instance the following table of the symmetric group of order 6 can be considered as the multiplication table with respect to the basis $\{A, B, C, D, E, F\}$ of a six dimensional algebra $\mathscr{B}$ over any field

	A	B	C	D	E	F
A	A	B	C	D	E	F
B	B	A	E	F	C	D
C	C	F	A	E	D	B
D	D	E	F	A	B	C
E	E	D	B	C	F	A
F	F	C	D	B	A	E

$$(13)$$

The associativity of table (13) is ensured by the fact that it is the operational table of the symmetric group of order 6.

As table (13) shows the algebra $\mathscr{B}$ is noncommutative. Now, if the elements P and Q of $\mathscr{B}$ are such that

$$P = 2A + B + C \qquad \text{and} \qquad Q = C - D + E + F$$

then in view of table (13) we have

$$PQ = A + B + 3C + 2E + F.$$

An algebra whose multiplication table is the operational table of a group is called a *group algebra*.

Let us also observe that since a semigroup (*see* page 5) is an associative algebraic system the operational table of a semigroup can be considered as a multiplication table of an algebra over any field.

For instance the following table of the semigroup mentioned on page 5 can be considered as the multiplication table with respect to the basis $\{C, D\}$ of a two dimensional algebra $\mathscr{C}$ over any field

	C	D
C	C	C
D	D	D

(14)

As table (14) shows the algebra $\mathscr{C}$ is noncommutative. Now, if the element P of $\mathscr{C}$ is such that

$$P = C + D$$

then in view of table (14) we have

$$P^2 = (C + D)^2 = 2C + 2D.$$

An algebra whose multiplication table is the operational table of a finite semigroup is called a *semigroup algebra*.

Exercises

1. Let a_{ij}^k be the multiplication constants with respect to the basis $\{A_1, A_2, \dots, A_n\}$ of an algebra $\mathscr{A}$. Express the associativity condition of multiplication in $\mathscr{A}$ solely in terms of a_{ij}^k. Do the same for the commutativity condition of multiplication in $\mathscr{A}$.

2. Prove that (1) and (2) are equivalent to (4).

3. Let $\{A\}$ be a basis of an algebra $\mathscr{A}$ over the field of rationals and let $A^2 = 2A$. Does $\mathscr{A}$ have an element I such that $IX = XI = X$ for every element X of $\mathscr{A}$? If so, determine I as a scalar multiple of A.

4. Consider the algebra $\mathscr{A}$ in Problem 3. Determine all the elements X of $\mathscr{A}$ such that $X^2 = X$.

5. Let $\{A, B\}$ be a basis of a vector space $\mathscr{A}$ over the field of rationals. Define

$$A^2 = AB = BA = B^2 = A.$$

Prove that $\mathscr{A}$ is an algebra with respect to the above multiplication table. Determine the multiplication table of $\mathscr{A}$ with respect to the basis $\{C, D\}$ if $C = A + B$ and $D = A - B$.

6. Let $\{A, B\}$ be a basis of a two dimensional algebra $\mathscr{A}$ over the field of rationals. If $A^2 = B^2 = 0$ prove that $\mathscr{A}$ has no element I such that $IX = XI = X$ for every element X of $\mathscr{A}$.

7. Let $\{A, A^2\}$ be a basis of a two dimensional algebra $\mathscr{A}$ over a field of characteristic $\neq 2$. Prove that $\mathscr{A}$ is commutative. Moreover, if $A^3 = A$ prove that $\mathscr{A}$ has a basis $\{C, D\}$ such that

$$C^2 = 2C, \qquad D^2 = 2D, \qquad CD = DC = 0.$$

8. Let $\mathscr{A}$ be a two dimensional algebra over the field of real numbers such that $X^2 = 0$ for every element X of $\mathscr{A}$. Prove that $XY = 0$ for every element X and Y of $\mathscr{A}$.

9. Let $\{A, B\}$ be a basis of a two dimensional algebra $\mathscr{A}$ over the field of rationals. If $A^2 = B$ and $AB = BA = B^2 = 0$ prove that $XYZ = 0$ for every element X, Y and Z of $\mathscr{A}$.

10. Let $\{A, B, C\}$ be a basis of a vector space $\mathscr{A}$. Define $AA = BB = A$, $AB = BA = B$, $AC = CA = BC = CB = C$, $CC = 0$. Prove that $\mathscr{A}$ is an algebra with respect to the above multiplication table.

3.2. Idempotent and Nilpotent Elements of an Algebra

Perhaps a typical situation in studying algebras is the following. It is required to prove the existence of an algebra $\mathscr{A}$ over a given field by determining (as completely as possible) a multiplication table of $\mathscr{A}$ (*i.e.*, a multiplication table for the elements of a basis of $\mathscr{A}$) provided $\mathscr{A}$ is assumed to satisfy certain conditions.

Let us emphasize that a multiplication table of an algebra $\mathscr{A}$ is a most significant single item which reveals various features of $\mathscr{A}$. Indeed, it is not too inappropriate to call a multiplication table of $\mathscr{A}$ the *structure* of $\mathscr{A}$.

For instance, let us prove the existence of a two dimensional algebra $\mathscr{A}$ over the field of rational numbers $\mathfrak{F}$, provided the basis elements A and B of $\mathscr{A}$ are assumed to satisfy the following conditions:

$$A^2 = A \qquad \text{and} \qquad B^2 = 0 \tag{15}$$

Naturally, to prove the existence of an algebra $\mathscr{A}$ over $\mathfrak{F}$ whose basis elements A and B are assumed to satisfy (15) it is enough to complete the following multiplication table

	A	B
A	A	$mA + nB$
B	$pA + qB$	0

(16)

i.e., to determine the elements m, n, p and q of $\tilde{\jmath}$ such that (16) becomes an *associative* table (*see* page 68).

Thus, for instance, we must have $(AB)B = A(BB)$ which in view of (16) implies

$$(AB)B = (mA + nB)B = mAB = A(BB) = 0$$

which is satisfied if we choose, say,

$$m = 0 \qquad (17)$$

In view of (17) we see that table (16) becomes

	A	B
A	A	nB
B	$pA + qB$	0

(18)

Again, we must have $(AA)B = A(AB)$ which in view of (18) implies

$$(AA)B = AB - nB = A(AB) = nAB = n^2B$$

which is satisfied if we choose, say,

$$n = 1. \qquad (19)$$

In view of (19) we see that table (18) becomes

	A	B
A	A	B
B	$pA + qB$	0

(20)

Similarly, we must have $(BB)A = B(BA)$ which in view of (20) implies

$$(BB)A = 0 = B(BA) = pBA$$

which is satisfied if we choose, say,

$$p = 0. \qquad (21)$$

Also, we must have $(BA)A = B(AA)$ which in view of (20) and (21) implies

$$(BA)A = qBA = q^2B = B(AA) = BA = qB$$

which is satisfied if we choose, say,

$$q = 1. \tag{22}$$

Thus, in view of (20), (21) and (22), multiplication table (16) becomes

$$
\begin{array}{c|cc}
 & A & B \\
\hline
A & A & B \\
B & B & 0
\end{array}
\tag{23}
$$

It is easy to verify that (23) is indeed a multiplication table of a two dimensional algebra whose basis elements A and B satisfy the conditions given in (15). The reader may also verify this by applying Light's associativity test.*

Let us mention also that table (23) is not the only multiplication table of a two dimensional algebra whose basis elements A and B satisfy the conditions given in (15). For instance, the table below

$$
\begin{array}{c|cc}
 & A & B \\
\hline
A & A & B \\
B & 0 & 0
\end{array}
\tag{24}
$$

also proves the existence of a two dimensional algebra satisfying the conditions given in (15).

One of the reasons which made the determination of m, n, p and q fairly convenient is the fact that algebra $\mathscr{A}$ had elements A and B such that

$$A^2 = A \quad \text{and} \quad B^2 = 0.$$

According to the terminology introduced on page 9, an element X of an algebra is called *idempotent* if $X^2 = X$. Moreover, an element Y of an algebra is called *nilpotent of index k* if $Y^k = 0$ and $Y^{k-1} \neq 0$ for some positive natural number k.

The above example shows that in determining a multiplication table of an algebra $\mathscr{A}$ it is very convenient to write the multiplication table of $\mathscr{A}$ with respect to a basis of $\mathscr{A}$ which includes as many as possible nonzero idempotent or nonzero nilpotent elements of $\mathscr{A}$.

In this connection, we have

LEMMA 2.   *Let N be a nilpotent element of index $k > 1$ of an algebra $\mathscr{A}$. Then*

$$N, \quad N^2, \ldots, N^{k-1}$$

are $k - 1$ linearly independent elements of $\mathscr{A}$.

---

*See Clifford, A. H., and Preston, G. B., *The Algebraic Theory of Semigroups*, Vol. 1, p. 7. American Mathematical Society Surveys No. 7. Providence, R. I. 1961.

Proof. Assume the contrary and let, without loss of generality,

$$N + sN^2 + \cdots + tN^{k-1} = 0$$

for some scalars $s, \ldots, t$. But then

$$N^{k-1} + sN^k + \cdots + tN^{2k-2} = 0.$$

However, since $k > 1$ and $N^k = 0$ the above equality implies $N^{k-1} = 0$ which contradicts the fact that $N^{k-1} \neq 0$.

Thus, Lemma 2 is proved.

Let $\mathscr{A}$ be an algebra over a field $\mathfrak{F}$. By an *n*th *degree polynomial* (in the present context) we mean an expression of the form $a_0 I$ if $a_n = 0$ for $n > 0$, otherwise

$$a_n X^n + \cdots + a_1 X + a_0 I \quad \text{with} \quad a_i \in \mathfrak{F} \quad \text{and} \quad a_n \neq 0 \quad \text{for} \quad n > 0 \tag{25}$$

if $\mathscr{A}$ has a unit element I; or else, we mean an expression of the form $a_1 X$ if $a_n = 0$ for $n > 1$, otherwise

$$a_n X^n + \cdots + a_1 X \quad \text{with} \quad a_i \in \mathfrak{F} \quad \text{and} \quad a_n \neq 0 \quad \text{for} \quad n > 1. \tag{26}$$

Let A be an element of an n dimensional algebra $\mathscr{A}$ over a field $\mathfrak{F}$. Since

$$A, \ A^2, \ldots, A^n, \ A^{n+1}$$

are linearly dependent it is clear that there exists a unique monic polynomial of least degree

$$a_m X^m + a_{m-1} X^{m-1} + \cdots \tag{27}$$

which may be of type (25) or (26) and of which A is a root, *i.e.*,

$$a_m A^m + a_{m-1} A^{m-1} + \cdots = 0. \tag{28}$$

As expected the polynomial given in (27) is called the *minimum polynomial* of A (*see* page 62). Moreover,

$$a_m X^m + a_{m-1} X^{m-1} + \cdots = 0$$

is called the *minimum equation* of A (*see* page 62).

In view of (28) it is customary to say that every element A of $\mathscr{A}$ satisfies its minimum equation.

It can be readily verified that the minimum equation of 0 is

$$X = 0 \tag{29}$$

and the minimum equation of the unit element I (if it exists) of an algebra is

$$X - I = 0. \tag{30}$$

LEMMA 3. *Let E be an element of an algebra $\mathscr{A}$. Then the minimum equation of E is*

$$X^2 - X = 0 \tag{31}$$

if and only if E is a nonzero idempotent element of $\mathscr{A}$ which is also different from the unit element I (if it exists) of $\mathscr{A}$.

Proof. If (31) is the minimum equation of E then clearly E is an idempotent element. Moreover, in view of (29) and (30) we see that $E \neq 0$ and $E \neq I$.

On the other hand, let E be an idempotent such that $E \neq 0$ and $E \neq I$. Then clearly, E satisfies (31) and it cannot satisfy any equation of degree less than 2 since this would contradict the fact that E is idempotent and $E \neq 0$ and $E \neq I$.

In view of (29), (30) and Lemma 3 we have

COROLLARY 1. *An element E of an algebra $\mathscr{A}$ is idempotent if and only if the minimum equation of E is*

$$X = 0 \qquad or \qquad X = I \qquad or \qquad X^2 = X$$

where I is the unit element (if it exists) of $\mathscr{A}$.

Next we prove

LEMMA 4. *Let N be an element of an algebra $\mathscr{A}$. Then the minimum equation of N is*

$$X^k = 0 \qquad with \qquad k > 0 \tag{32}$$

if and only if N is a nilpotent element of index k of $\mathscr{A}$.

Proof. If (32) is the minimum equation of N then N is a nilpotent element of index $N^{k-1} \neq 0$.

On the other hand, let N be a nilpotent element of index $k > 0$. Then clearly N satisfies (32) and cannot satisfy an equation of lower degree since (as in the proof of Lemma 2) this would contradict the fact that $N^{k-1} \neq 0$.

Below, we obtain some results concerning the existence of nonzero idempotent and nilpotent elements of an algebra. In particular, Theorem 1 gives a necessary and sufficient condition for the existence of a nonzero idempotent element in an algebra.

LEMMA 5. *Let the minimum equation of a nonzero element A of an algebra $\mathscr{A}$ be*

$$X^m + \cdots + a_2 X^2 + a_1 X = 0 \qquad \text{with} \qquad a_1 \neq 0. \tag{33}$$

Then $\mathscr{A}$ has a nonzero idempotent element E where

$$E = a_1^{-1}(A^{m-1} + \cdots + a_2 A). \tag{34}$$

Moreover, E is not the unit element of $\mathscr{A}$.

Proof. Since (33) is the minimum equation of a nonzero element A we see that $m > 1$ and

$$(A^{m-1} + \cdots + a_2 A)A = -a_1 A. \tag{35}$$

From (35) we obtain the following $m - 1$ equalities

$$(A^{m-1} + \cdots + a_2 A)(a_2 A) = -a_1(a_2 A)$$
$$\cdots \qquad \cdots \qquad = \cdots$$
$$(A^{m-1} + \cdots + a_2 A)(A^{m-1}) = -a_1(A^{m-1}).$$

Adding the left as well as the right sides of the above $m - 1$ equalities we obtain

$$(A^{m-1} + \cdots + a_2 A)(A^{m-1} + \cdots + a_2 A) = -a_1(A^{m-1} + \cdots + a_2 A)$$

from which it follows that

$$(-a_1^{-1}(A^{m-1} + \cdots + a_2 A))^2 = -a_1^{-1}(A^{m-1} + \cdots + a_2 A)$$

which in view of (34) implies $E^2 = E$. Thus, E is an idempotent element of $\mathscr{A}$.

Moreover, $E \neq 0$ and $E \neq I$ since otherwise, as (34) shows, the minimum equation of A would be of degree less than m, contradicting (33).

LEMMA 6. *Let the minimum equation of a nonnilpotent element B of an algebra $\mathscr{A}$ be*

$$X^m + \cdots + b_u X^u + b_k X^k = 0 \qquad \text{with} \qquad k \geq 2 \tag{36}$$

where $b_i \neq 0$ for $i = k, u, \ldots$.

Then $\mathscr{A}$ has a nilpotent element of index 2 and a nonzero idempotent element which is not the unit element of $\mathscr{A}$.

Proof. Since (36) is the minimum equation of a nonnilpotent element B we see that $m > k$ and

$$(B^{m-1} + \cdots + b_u B^{u-1} + b_k B^{k-1})B = 0. \tag{37}$$

75

From (37) we obtain the following $m - k + 1$ equalities

$$(B^{m-1} + \cdots + b_u B^{u-1} + b_k B^{k-1})(b_k B^{k-1}) = 0$$
$$\cdots \qquad\qquad\qquad \cdots \qquad\qquad = 0$$
$$(B^{m-1} + \cdots + b_u B^{u-1} + b_k B^{k-1})(B^{m-1}) \quad = 0.$$

Adding the left as well as the right sides of the above $m - k + 1$ equalities, we obtain

$$(B^{m-1} + \cdots + b_u B^{u-1} + b_k B^{k-1})(B^{m-1} + \cdots + b_u B^{u-1} + b_k B^{k-1}) = 0$$

which shows that

$$N = B^{m-1} + \cdots + b_u B^{u-1} + b_k B^{k-1} \tag{38}$$

is such that $N^2 = 0$. Obviously, $N \neq 0$ since otherwise, as (38) shows, the minimum equation of B would be of degree less than m, contradicting (36).

Thus, indeed N is a nilpotent element of index 2 of $\mathscr{A}$.

To prove the existence of a nonzero idempotent element of $\mathscr{A}$, let us observe that (36) yields equation

$$b_n X^n + \cdots + b_h X^h + b_k X^k = 0 \tag{39}$$

where $h \geqslant \max\{2k, u\}$ and $m + k > n \geqslant h$ and $b_n \neq 0$.

To see this let us consider an example. Let (36) be

$$X^5 + X^4 - X^3 = 0. \tag{40}$$

Multiplying both sides of the above by X we have

$$X^6 + X^5 - X^4 = 0. \tag{41}$$

Adding (40) to (41) we obtain

$$X^6 + 2X^5 - X^3 = 0. \tag{42}$$

Multiplying both sides of the above by $2X^2$ we have

$$2X^8 + 4X^7 - 2X^5 = 0. \tag{43}$$

Adding (42) to (43) we obtain

$$2X^8 + 4X^7 + X^6 - X^3 = 0. \tag{44}$$

But from (40) it follows that

$$-2X^8 - 2X^7 + 2X^6 = 0. \tag{45}$$

Adding (44) to (45) we obtain

$$2X^7 + 3X^6 - X^3 = 0 \tag{46}$$

as desired.

Continuing with our proof, from (39) it follows that

$$b_n B^n + \cdots + b_h B^h = - b_k B^k$$

and hence

$$(b_n B^{n-k} + \cdots + b_h B^{h-k}) B^k = - b_k B^k \tag{47}$$

where $n - k \geqslant k$ since $n \geqslant h \geqslant \max\{2k, u\}$.

From (47) we obtain the following $n - h + 1$ equalities

$$(b_n B^{n-k} + \cdots + b_h B^{h-k})(b_h B^{h-k}) = - b_k(b_h B^{h-k})$$
$$\cdots \qquad \cdots \qquad = \qquad \cdots$$
$$(b_n B^{n-k} + \cdots + b_h B^{h-k})(b_n B^{n-k}) = - b_k(b_n B^{n-k}).$$

Adding the left as well as the right hand of the above $n - h + 1$ equalities, we obtain

$$(b_n B^{h-k} + \cdots + b_h B^{h-k})(b_n B^{n-k} + \cdots + b_h B^{h-k})$$
$$= - b_k(b_n B^{n-k} + \cdots + b_h B^{h-k})$$

from which it follows that

$$(- b_k^{-1}(b_n B^{n-k} + \cdots + b_h B^{h-k}))^2 = - b_k^{-1}(b_n B^{n-k} + \cdots + b_h B^{h-k})$$

which shows that

$$E = - b_k^{-1}(b_n B^{n-k} + \cdots + b_h B^{h-k}) \tag{48}$$

is such that $E^2 = E$. However, since $b_n \neq 0$ and $m + k > n$ we see that $E \neq 0$ and $E \neq I$ since otherwise, as (48) shows, the minimum equation of B would be of degree less than m, contradicting (36).

Thus, indeed E is a nonzero idempotent element of $\mathscr{A}$ and E is not the unit element of $\mathscr{A}$.

For instance, if the minimum equation of B is as given by (40) then in view of (46) we have

$$2B^7 + 3B^6 = B^3$$

and therefore

$$(2B^4 + 3B^3)B^3 = B^3$$

which yields

$$(2B^4 + 3B^3)(3B^3) = 3B^3$$

and

$$(2B^4 + 3B^3)(2B^4) = 2B^4.$$

Adding the above two equalities we see that

$$2B^4 + 3B^3$$

is a nonzero idempotent element which is not the unity of the algebra under consideration.

THEOREM 1. *Let $\mathscr{A}$ be an algebra. Then $\mathscr{A}$ has at least one nonzero idempotent element if and only if $\mathscr{A}$ has at least one nonnilpotent element.*

Proof. If E is a nonzero idempotent element of $\mathscr{A}$ then clearly E is a nonnilpotent element of $\mathscr{A}$.

Next, let C be a nonnilpotent element of $\mathscr{A}$. If $\mathscr{A}$ has a unit element then the latter is a nonzero idempotent element of $\mathscr{A}$. If $\mathscr{A}$ has no unit element then clearly the minimum equation of C is of the form given by (33) or by (36). But then Lemmas 5 and 6 respectively ensure the existence of a nonzero idempotent element of $\mathscr{A}$.

In view of Theorem 1 we have:

COROLLARY 2. *An algebra has no nonzero idempotent element if and only if every element of the algebra is nilpotent.*

As expected, an element A of an algebra $\mathscr{A}$ is called *nonsingular* or *regular* (*see* page 9) if A has a multiplicative inverse A^{-1} in $\mathscr{A}$. Naturally, $AA^{-1} = A^{-1}A = I$ where I is the unit element of $\mathscr{A}$.

Moreover, a nonzero element A of an algebra $\mathscr{A}$ is called a *divisor of zero* (*see* page 9) if $AB = 0$ or $CA = 0$ for some nonzero element B or C of $\mathscr{A}$.

Clearly, if A is nonsingular it cannot be a divisor of zero.

As in the case of matrices (*see* page 62), here also we have

LEMMA 7. *Let $\mathscr{A}$ be an algebra with unity and let*

$$X^m + \cdots + a_1 X + a_0 I = 0 \qquad with \qquad m \geq 1 \qquad (49)$$

be the minimum equation of an element A of $\mathscr{A}$.
Then A is nonsingular if and only if $a_0 \neq 0$.

Proof. Let A be nonsingular. Assume $a_0 = 0$. But then $(A^{m-1} + \cdots + a_1 I)A = 0$ implying that A is a divisor of zero since, as (49) shows, the minimum equation of A is of degree m. Thus, $a_0 \neq 0$.

Next, let $a_0 \neq 0$. But then from (49) it follows that

$$-a_0^{-1}(A^{m-1} + \cdots + a_1 I)A = I \qquad (50)$$

showing that $-a_0^{-1}(A^{m-1} + \cdots + a_1 I)$ is the multiplicative inverse of A. Thus, A is nonsingular.

Incidentally, (50) gives the multiplicative inverse of A as a linear combination of I and powers of A.

Exercises

1. Prove that there exists a two dimensional algebra with basis $\{A, B\}$ such that

$$A^3 = 0 \qquad \text{and} \qquad A^2 = B.$$

2. Prove that there exists a three dimensional algebra with basis $\{A, B, C\}$ such that

$$AA = A, \qquad AB = BA = B, \qquad AC = CA = BB = C$$
$$BC = CB = CC = 0.$$

3. Prove that there exists a three dimensional algebra with basis $\{A, B, C\}$ such that

$$AA = B, \qquad AC = B = -CA$$
$$AB = BA = BB = BC = CB = CC = 0.$$

4. Prove that there exists a three dimensional algebra with basis $\{A, B, C\}$ such that

$$AA = B, \qquad AB = BA = C$$
$$BB = AC = CA = BC = CB = CC = 0.$$

Prove also that this algebra has no nonzero idempotent element.

5. Let the minimum equation of an element A of an algebra $\mathscr{A}$ over the field of rational numbers be

$$X^5 + 3X^3 = 0.$$

Prove that $\mathscr{A}$ has a nilpotent element N of index 2 and a nonzero idempotent element E. Express N as well as E as a linear combination of powers of A.

6. Determine whether or not there exists a noncommutative four dimensional algebra $\mathscr{A}$ over the field of rational numbers such that $\mathscr{A}$ has a basis with two idempotent and two nilpotent elements of index 2.

7. Let $\mathscr{A}$ be a three dimensional algebra with unity over the field of real numbers. Prove that $\mathscr{A}$ has at least one singular (*i.e.*, which has no multiplicative inverse) element.

8. Let $\mathscr{A}$ be an algebra without unity. Let $X^k = X$ for $k \geqq 2$ be the minimum equation of an element of $\mathscr{A}$. Prove that for every natural number $n \geqq 2$ there exists a nonnilpotent element of $\mathscr{A}$ satisfying the equation $X^n = X$.

9. Let $\mathscr{A}$ be a two dimensional algebra having no divisor of zero. Prove that $\mathscr{A}$ has unity and that every nonzero element of $\mathscr{A}$ has a multiplicative inverse.

10. Let $\mathscr{A}$ be a two dimensional algebra over the field of integers modulo 3. Prove that $\mathscr{A}$ is commutative, has unity and that every nonzero element of $\mathscr{A}$ is regular (*i.e.*, has a multiplicative inverse).

3.3. Ideals of an Algebra

In this section we shall be concerned with subsets of algebras. Since an algebra is a ring, many of the definitions, terminologies and results which were introduced and obtained in connection with the subsets of a ring (*see* Sections 1.1 and 1.2, pages 10 and 16) will be used in connection with the subsets of an algebra. Similarly, since an algebra is a vector space, many of the definitions, terminologies and results which were introduced and obtained in connection with the subsets of a vector space (*see* Sections 1.1 and 1.2, pages 42 and 51) will also be used in connection with the subsets of an algebra.

Let $\mathscr{S}_1$ and $\mathscr{S}_2$ be subsets of an algebra $\mathscr{A}$. Then as in (18) on page 16, we define the *sum*

$$\mathscr{S}_1 + \mathscr{S}_2 \tag{51}$$

as the set of all sums $A_1 + A_2$ with $A_1 \in \mathscr{S}_1$ and $A_2 \in \mathscr{S}_2$. Clearly, *addition of subsets* of $\mathscr{A}$ is both an associative and commutative operation.

As expected, for finitely many subsets $\mathscr{S}_i$ of $\mathscr{A}$, we define the sum

$$\mathscr{S}_1 + \mathscr{S}_2 + \cdots + \mathscr{S}_n = \sum_i \mathscr{S}_i \tag{52}$$

(by induction) as the set of all sums $A_1 + A_2 + \cdots + A_n = \sum A_i$ where $A_i \in \mathscr{S}_i$.

Moreover, as in (19) on page 16, for the subsets $\mathscr{S}_1$ and $\mathscr{S}_2$ of an algebra $\mathscr{A}$ we define the *product*

$$\mathscr{S}_1 \mathscr{S}_2 \tag{53}$$

as the set of all finite sums of products $A_1 A_2$ with $A_1 \in \mathscr{S}_1$ and $A_2 \in \mathscr{S}_2$. Clearly, *multiplication of subsets* of $\mathscr{A}$ is an associative (but not necessarily commutative) operation.

Again, as expected, for finitely many subsets of $\mathscr{S}_i$ of $\mathscr{A}$ we define the product

$$\mathscr{S}_1 \mathscr{S}_2 \cdots \mathscr{S}_n = \prod_i \mathscr{S}_i \tag{54}$$

(by induction) as the set of all finite sums of products $A_1 A_2 \cdots A_n = \Pi A_i$ where $A_i \in \mathscr{S}_i$.

In particular, for every subset $\mathscr{S}$ of $\mathscr{A}$ and every element A of $\mathscr{A}$, the notations

$$\mathscr{S}A \quad \text{and} \quad A\mathscr{S} \tag{55}$$

are used instead of $\mathscr{S}\{A\}$ and $\{A\}\mathscr{S}$ respectively. Clearly, as mentioned in connection with (20) on page 16, we see that $\mathscr{S}A$ is the set of all the elements of $\mathscr{A}$ of the form SA with $S \in \mathscr{S}$. Similarly, $A\mathscr{S}$ is the set of all the elements of $\mathscr{A}$ of the form SA with $S \in \mathscr{S}$.

If in the product given by (54) every factor $\mathscr{S}_i$ is equal to a subset $\mathscr{S}$

of $\mathscr{A}$ then the product is denoted by

$$\mathscr{S}^n.$$

Thus,

$$\prod_{i=1}^{n} \mathscr{S}_i = \mathscr{S}^n \qquad \text{if } \mathscr{S}_i = \mathscr{S} \qquad (i = 1, 2, \ldots, n). \tag{56}$$

Finally, the intersection

$$\mathscr{S}_1 \cap \mathscr{S}_2 \tag{57}$$

of the subsets $\mathscr{S}_1$ and $\mathscr{S}_2$ of $\mathscr{A}$ is defined, as usual, as the set of all elements which are common to both $\mathscr{S}_1$ and $\mathscr{S}_2$. Obviously, *intersection of subsets* of $\mathscr{A}$ is both an associative and commutative operation. Moreover, the notation

$$\bigcap_{i} \mathscr{S}_i \tag{58}$$

for finitely or infinitely many subsets $\mathscr{S}_i$ of $\mathscr{A}$ is self-explanatory.

Various distributivity formulas which are given by (21), (22) and (23) on page 17 hold in connection with the subsets of an algebra.

A subset $\mathscr{S}$ of an algebra $\mathscr{A}$ over a field $\mathfrak{F}$ is called a *subspace* of $\mathscr{A}$ if $\mathscr{S}$ is a subspace of the vector space $\mathscr{A}$ (*see* page 41). Thus, according to (8) on page 41 a nonempty subset $\mathscr{S}$ of $\mathscr{A}$ is a subspace of $\mathscr{A}$ if and only if

$$(X - Y) \in \mathscr{S} \qquad \text{and} \qquad sX \in \mathscr{S} \tag{59}$$

for every element X and Y of $\mathscr{A}$ and every element s of $\mathfrak{F}$.

In view of (52), (54), (58) and (59) we see that if $\mathscr{S}_i$ for $i = 1, 2, \ldots, n$ is a subspace of an algebra $\mathscr{A}$ then each of

$$\sum \mathscr{S}_i, \qquad \Pi \mathscr{S}_i, \qquad \cap \mathscr{S}_i$$

is a subspace of $\mathscr{A}$.

If $\mathscr{S} = \{A_1, A_2, \ldots, A_n\}$ is a subset of an algebra $\mathscr{A}$ then as indicated on page 42, we denote by

$$[\mathscr{S}] \qquad \text{or} \qquad [\{A_1, A_2, \ldots, A_n\}] \tag{60}$$

the subspace generated or spanned by $\mathscr{S}$, *i.e.*, the intersection of all the subspaces of $\mathscr{A}$ each containing $\mathscr{S}$. Usually, (60) is simply denoted by

$$[A_1, A_2, \ldots, A_n]. \tag{61}$$

If $\{U_1, U_2, \ldots, U_m\}$ is a basis of a subspace $\mathscr{S}_1$ of an algebra $\mathscr{A}$ and $\{V_1, V_2, \ldots, V_n\}$ is a basis of a subspace $\mathscr{S}_2$ of $\mathscr{A}$ then the subspace $\mathscr{S}_1\mathscr{S}_2$ is generated by $\{U_1V_1, U_1V_2, \ldots, U_mV_n\}$. Thus,

$$\dim \mathscr{S}_1\mathscr{S}_2 \leqslant (\dim \mathscr{S}_1)(\dim \mathscr{S}_2) = mn. \tag{62}$$

Obviously, in (62) equality holds if and only if $U_i V_j$ are linearly independent for $i = 1, 2, \ldots, m$ and $j = 1, 2, \ldots, n$. Thus, it is easily seen that we have:

LEMMA 8. *Let $\{U_1, U_2, \ldots, U_m\}$ be a basis of a subspace $\mathscr{S}_1$ of an algebra $\mathscr{A}$ and $\mathscr{S}_2$ be an n dimensional subspace of $\mathscr{A}$. Then*

$$\dim \mathscr{S}_1 \mathscr{S}_2 = mn$$

if and only if $U_1 A_1 + U_2 A_2 + \cdots + U_m A_m \neq 0$ where A_i's are elements of $\mathscr{S}_2$ and not all zero.

Using notation (55) we have the following

COROLLARY 3. *Let $\mathscr{S}$ be a subspace of an algebra $\mathscr{A}$ and A an element of $\mathscr{A}$. Then*

$$\dim A\mathscr{S} = \dim \mathscr{S}$$

(or $\dim \mathscr{S}A = \dim \mathscr{S}$) if and only if

$$AS \neq 0$$

(or $SA \neq 0$) for every nonzero element S of $\mathscr{S}$.

A subspace $\mathscr{S}$ of an algebra $\mathscr{A}$ over a field $\mathfrak{F}$ is called a subalgebra of $\mathscr{A}$ if $\mathscr{S}$ is an algebra over $\mathfrak{F}$ with respect to the operations in $\mathscr{A}$. Clearly, the subspace $\mathscr{S}$ is a subalgebra if and only if $XY \in \mathscr{S}$ for every element X and Y of $\mathscr{S}$.

Let us observe that if $\mathscr{S}$ is a subalgebra of an algebra $\mathscr{A}$ then $\mathscr{S}$ is a subring of the ring $\mathscr{A}$.

LEMMA 9. *Let $\mathscr{S}$ be a subspace of an algebra $\mathscr{A}$. Then $\mathscr{S}$ is a subalgebra of $\mathscr{A}$ if and only if*

$$\mathscr{S}^2 \subset \mathscr{S}.$$

Proof. If $\mathscr{S}$ is a subalgebra then clearly any finite sum of products XY with X and Y elements of $\mathscr{S}$ is again an element of $\mathscr{S}$. Thus, in view of (53) we see that $\mathscr{S}^2 \subset \mathscr{S}$.

Next, let $\mathscr{S}$ be a subspace of $\mathscr{A}$. If $\mathscr{S}^2 \subset \mathscr{S}$ then clearly the product XY of the elements X and Y of $\mathscr{S}$ is again an element of $\mathscr{S}$ implying that $\mathscr{S}$ is a subalgebra of $\mathscr{A}$.

It is readily seen that if $\mathscr{A}$ is an algebra then

$$\mathscr{A} = \mathscr{A}^1 \supset \mathscr{A}^2 \supset \mathscr{A}^3 \supset \cdots \qquad (63)$$

where $\mathscr{A}^k$ is a subalgebra of $\mathscr{A}^h$ for $k \geqslant h$. However, since $\mathscr{A}$ is of finite dimension we see that in (63) we must have $\mathscr{A}^s = \mathscr{A}^{s+1}$ for some natural number s and, consequently, $\mathscr{A}^n = \mathscr{A}^s$ for every $n \geqslant s$.

Clearly, if an algebra $\mathscr{A}$ has a unity then

$$\mathscr{A}^2 = \mathscr{A}. \tag{64}$$

Let us observe that if $\mathscr{S}_1$ and $\mathscr{S}_2$ are subalgebras of an algebra $\mathscr{A}$ then, in general, neither their sum, as given by (51), nor their product, as given by (53), is a subalgebra of $\mathscr{A}$. However, their intersection, as given by (57), is a subalgebra of $\mathscr{A}$. Moreover, if $\mathscr{S}$ is a nonempty subset of $\mathscr{A}$ then the intersection of all subalgebras of $\mathscr{A}$ each containing $\mathscr{S}$ is a subalgebra of $\mathscr{A}$ which is called the *subalgebra generated* by $\mathscr{S}$ and is denoted by $(\mathscr{S})$. Usually, the subalgebra generated by a singleton $\{A\}$ is simply called the subalgebra generated by A and is denoted by (A).

Let A and B be elements of an algebra $\mathscr{A}$. Then, based on the notation given by (55) and (53), we see that each of

$$A\mathscr{A}, \qquad \mathscr{A}B, \qquad A\mathscr{A}B, \qquad \mathscr{A}B\mathscr{A} \tag{65}$$

is a subalgebra of $\mathscr{A}$. In (65), in view of (53) and (55) the subalgebra $\mathscr{A}B\mathscr{A}$ is the set of all finite sums of summands each of the form XAY with X and Y elements of $\mathscr{A}$ (*see* page 11).

Below we give some more examples of subalgebras.

Let $\mathscr{S}$ be a nonempty subset of an algebra $\mathscr{A}$. Then the set

$$\mathrm{czr}(\mathscr{S})$$

of all the elements X of $\mathscr{A}$ such that $XS = SX$ for every element S of $\mathscr{S}$ is easily seen to form a subalgebra of $\mathscr{A}$ and is called the *centralizer* of $\mathscr{S}$. In particular, the subalgebra

$$\mathrm{czr}\,(\mathscr{A}) = \mathrm{cnt}\,(\mathscr{A})$$

of an algebra $\mathscr{A}$ is called the *center* of $\mathscr{A}$ (*see* page 10).

Let us observe that the center of an algebra $\mathscr{A}$ is a commutative subalgebra of $\mathscr{A}$.

Let $\mathscr{A}$ be an algebra over a field $\mathfrak{F}$ and let $\mathscr{Z}$ be the center of $\mathscr{A}$. If $\mathscr{A}$ has a unit element I then obviously

$$(I) \subset \mathscr{Z} \tag{66}$$

where (I) is the subalgebra of $\mathscr{A}$ generated by I (*i.e.*, the set of all products sI with $s \in \mathfrak{F}$).

If in (66) the equality holds, then $\mathscr{A}$ is called a central algebra over the field $\mathfrak{F}$. Thus, we introduce

DEFINITION 2. *An algebra $\mathscr{A}$ with a unit element I over a field $\mathfrak{F}$ is called central over $\mathfrak{F}$ if the center of $\mathscr{A}$ is equal to the subalgebra of $\mathscr{A}$ generated by I.*

For instance, if $\{A\}$ is a basis of a one dimensional algebra $\mathscr{A}$ over a field $\mathfrak{F}$ and if $AA = A$ then $\mathscr{A}$ is central over $\mathfrak{F}$. On the other hand, the two dimensional algebra $\mathscr{A}$ whose multiplication table is given by (23) is not central over the field of rational numbers.

Let $E_1, E_2, \ldots, E_n$ be elements of an algebra $\mathscr{A}$. We say that $E_1, E_2, \ldots, E_n$ are *pairwise orthogonal idempotents* of $\mathscr{A}$ if

$$E_i^2 = E_i \quad \text{and} \quad E_i E_j = 0. \quad (i \neq j, \quad i,j = 1, 2, \ldots, n) \quad (67)$$

Clearly, a set $\{E_1, E_2, \ldots, E_n\}$ of nonzero pairwise orthogonal idempotent elements of an algebra is a linearly independent subset of $\mathscr{A}$ (*see* Definition 2, page 42). Because if

$$s_1 E_1 + \cdots + s_i E_i + \cdots + s_n E_n = 0$$

then

$$s_1 E_1 E_i + \cdots + s_i E_i E_i + \cdots + s_n E_n E_i = 0$$

which in view of (67) implies that $s_i = 0$.

LEMMA 10. *Let* $[E_1, E_2, \ldots, E_n]$ *be the subspace of an algebra* $\mathscr{A}$ *spanned by the n pairwise orthogonal nonzero idempotent elements* $E_1, E_2, \ldots, E_n$ *of* $\mathscr{A}$. *Then* $\mathscr{B} = [E_1, E_2, \ldots, E_n]$ *is an n dimensional subalgebra of* $\mathscr{A}$ *and* $E = E_1 + E_2 + \cdots + E_n$ *is the unity of* $\mathscr{B}$.

Proof. In view of (67) we see that

$$(\textstyle\sum s_i E_i)(\sum t_i E_i) \in [E_1, E_2, \ldots, E_n]$$

and

$$(\textstyle\sum E_i)(\sum s_i E_i) = (\sum s_i E_i)(\sum E_i) = \sum s_i E_i.$$

Thus, indeed $\mathscr{B}$ is a subalgebra of $\mathscr{A}$ and $E = \sum E_i$ is the unity of $\mathscr{B}$. Moreover, since $E_1, E_2, \ldots, E_n$ are linearly independent, we see that dim $\mathscr{B} = n$, as desired.

Motivated by the definition of an ideal of a ring (*see* page 11), a subspace $\mathscr{I}$ of an algebra $\mathscr{A}$ is called a *two sided ideal* or simply an *ideal* of $\mathscr{A}$ if

$$AB \in \mathscr{I} \quad \text{and} \quad BA \in \mathscr{I} \quad (68)$$

for every element B of $\mathscr{I}$ and every element A of $\mathscr{A}$.

If the first membership relation in (68) holds, then $\mathscr{I}$ is called a *left ideal* of $\mathscr{A}$ and if the second holds then $\mathscr{I}$ is called a *right ideal* of $\mathscr{A}$. These ideals are called *one-sided ideals* of $\mathscr{A}$.

From (68) it follows that a subspace $\mathscr{I}$ of an algebra $\mathscr{A}$ is an ideal of $\mathscr{A}$ if and only if

$$\mathscr{A}\mathscr{I} \subset \mathscr{I} \quad \text{and} \quad \mathscr{I}\mathscr{A} \subset \mathscr{I}. \quad (69)$$

Moreover, $\mathscr{S}$ is a left (right) ideal of $\mathscr{A}$ if and only if the first (second) inclusion in (69) holds.

In view of Lemma 9 we see that any one-sided ideal of an algebra $\mathscr{A}$ is a subalgebra of $\mathscr{A}$.

Let $\mathscr{A}$ be an algebra. Then each of

$$\{0\}, \quad \mathscr{A}, \quad \mathscr{A}^2, \quad \mathscr{A}^3, \ldots \tag{70}$$

is an ideal of $\mathscr{A}$.

As expected, the *zero ideal* $\{0\}$ and $\mathscr{A}$ are called the *trivial* ideals of $\mathscr{A}$ and any other ideal of $\mathscr{A}$ is called *nontrivial*.

If $\mathscr{S}$ is any subspace of an algebra $\mathscr{A}$ and $\mathscr{L}$ and $\mathscr{R}$ are respectively a left and a right ideal of $\mathscr{A}$ then, clearly,

$$\mathscr{L}\mathscr{S}, \quad \mathscr{S}\mathscr{R}, \quad \mathscr{L}\mathscr{R} \tag{71}$$

are respectively a left, a right and a two sided ideal of $\mathscr{A}$.

If $\mathscr{L}_i$ for $i = 1, 2, \ldots, n$ is a left ideal of an algebra $\mathscr{A}$ then each of

$$\prod_i \mathscr{L}_i, \quad \bigcap_i \mathscr{L}_i, \quad \sum_i \mathscr{L}_i \tag{72}$$

is a left ideal of $\mathscr{A}$.

Similarly, if $\mathscr{R}_i$ for $i = 1, 2, \ldots, n$ is a right ideal of an algebra $\mathscr{A}$ then each of

$$\prod_i \mathscr{R}_i, \quad \bigcap_i \mathscr{R}_i, \quad \sum_i \mathscr{R}_i$$

is a right ideal of $\mathscr{A}$.

Let us observe that for $i = 1, 2, \ldots, n$ inclusions (24) on page 17, hold for ideals $\mathscr{I}_i$ of an algebra $\mathscr{A}$. Moreover, from (71) it follows that $\mathscr{A}\mathscr{S}\mathscr{A}$ as well as $\mathscr{A}A\mathscr{A}$ is an ideal of $\mathscr{A}$ for every subspace $\mathscr{S}$ and every element A of $\mathscr{A}$.

Obviously, the intersection of any family of ideals of an algebra $\mathscr{A}$ is an ideal of $\mathscr{A}$. As expected (*see* page 11) the intersection of all (left, right) ideals of $\mathscr{A}$ each containing a nonempty subset $\mathscr{E}$ of $\mathscr{A}$ is called the (left, right) *ideal generated* by $\mathscr{E}$. Clearly, the ideal generated by $\mathscr{E}$ is the smallest (with respect to the set-theoretical inclusion) ideal of $\mathscr{A}$ containing $\mathscr{E}$.

As we shall see later, the existence or nonexistence of nontrivial ideals in an algebra $\mathscr{A}$ is of significant relevance as far as the structure of $\mathscr{A}$ is concerned. Obviously, if $\mathscr{A}$ is a *zero algebra*, i.e., $XY = 0$ for every element X and Y of $\mathscr{A}$ (*see* page 8) then every subspace of $\mathscr{A}$ is an ideal of $\mathscr{A}$. On the other hand, a one dimensional algebra has no nontrivial ideals and as such (excluding the zero algebra case) is a simple algebra (*see* page 11) according to the following definition.

DEFINITION 3. *An algebra $\mathcal{A}$ is called simple if $\mathcal{A}$ has no nontrivial ideal and if $\mathcal{A}$ is not a zero algebra.*

Let $\mathcal{A}$ and $\mathcal{B}$ be algebras over the same field $\mathfrak{F}$. Then a mapping φ from $\mathcal{A}$ into $\mathcal{B}$ is called a *homomorphism* (or, more precisely, an *algebra-homomorphism*) from $\mathcal{A}$ into $\mathcal{B}$ if

$$\varphi(X+Y) = \varphi(X) + \varphi(Y) \tag{73}$$

$$\varphi(XY) = \varphi(X)\varphi(Y) \tag{74}$$

$$\varphi(sX) = s\varphi(X) \tag{75}$$

for every element X and Y of $\mathcal{A}$ and every element s of $\mathfrak{F}$.

Let us observe that (73) and (75) state that φ is a homomorphism from the vector space $\mathcal{A}$ into the vector space $\mathcal{B}$ (*see* page 46). Moreover, (73) and (74) state that φ is a homomorphism from the ring $\mathcal{A}$ into the ring $\mathcal{B}$. Clearly, if φ is a homomorphism from an algebra $\mathcal{A}$ into an algebra $\mathcal{B}$ then the image of $\mathcal{A}$ under φ is a subalgebra of $\mathcal{B}$.

If φ is a homomorphism from an algebra $\mathcal{A}$ into an algebra $\mathcal{B}$ and if φ is one-to-one then, as expected, φ is called an *isomorphism from $\mathcal{A}$ into $\mathcal{B}$* (*see* page 46).

If φ is a homomorphism from an algebra $\mathcal{A}$ into $\mathcal{A}$ then φ is called an *endomorphism* of $\mathcal{A}$ (*see* page 55). As usual, an isomorphism from an algebra $\mathcal{A}$ onto $\mathcal{A}$ is called an *automorphism* of $\mathcal{A}$ (*see* page 55).

For instance, if A is an element of an algebra $\mathcal{A}$ and if A has a multiplicative inverse A^{-1} then $\varphi(X) = AXA^{-1}$ is an automorphism of $\mathcal{A}$. Such an automorphism of $\mathcal{A}$ is called an *inner automorphism* of $\mathcal{A}$.

If there exists an isomorphism φ from an algebra $\mathcal{A}$ onto an algebra $\mathcal{B}$ then (*see* page 46) we denote this by $A \cong B$ and we say that $\mathcal{A}$ and $\mathcal{B}$ are isomorphic images of each other, or simply, $\mathcal{A}$ and $\mathcal{B}$ are isomorphic.

Clearly, isomorphism is an equivalence relation in any set of algebras over the same field. Moreover, two isomorphic algebras are algebraically indistinguishable.

THEOREM 2. *Let $\mathcal{A}$ be a one dimensional algebra over a field $\mathfrak{F}$. Then $\mathcal{A}$ is a one dimensional zero algebra or is a field isomorphic to $\mathfrak{F}$.*

Proof. Let $\{A\}$ be a basis of $\mathcal{A}$ and let the multiplicative table of $\mathcal{A}$ be given by

$$AA = sA \qquad \text{with} \qquad s \in \mathfrak{F}.$$

If $s = 0$ then clearly $\mathcal{A}$ is a one dimensional zero algebra (*see* page 85).

If $s \neq 0$ then the mapping φ given by $\varphi(x) = (xs^{-1})A$ is an isomorphism from the field $\mathfrak{F}$ onto $\mathcal{A}$.

Clearly, a one dimensional zero algebra cannot be isomorphic to a field since, for instance, the former has no unit element whereas the latter has a unit element. Thus, in view of Theorem 2 we may say that there are two essentially distinct one dimensional algebras over a given field.

Let $\mathscr{I}$ be an ideal of an algebra $\mathscr{A}$ over a field $\mathfrak{F}$. Then it is readily seen that the relation

$$X = Y \,(\mathrm{mod}\,\mathscr{I})$$

defined by

$$X = Y \,(\mathrm{mod}\,\mathscr{I}) \qquad \text{if} \qquad (X - Y) \in \mathscr{I} \tag{76}$$

is an equivalence relation in $\mathscr{A}$. As usual we denote by

$$[X]$$

the equivalence class to which an element X of $\mathscr{A}$ belongs. Moreover, we denote by

$$\mathscr{A}/\mathscr{I} \tag{77}$$

the set of all equivalence classes $[X]$. It is easy to verify that $\mathscr{A}/\mathscr{I}$ is an algebra over $\mathfrak{F}$ if we define

$$\begin{aligned} [X] + [Y] &= [X + Y] \\ [X][Y] &= [XY] \\ s[X] &= [sX] \end{aligned} \tag{78}$$

for every element X and Y of $\mathscr{A}$ and every element s of $\mathfrak{F}$. The algebra $\mathscr{A}/\mathscr{I}$ is called the *quotient algebra* of $\mathscr{A}$ with respect to the ideal $\mathscr{I}$, or, the *difference algebra* of $\mathscr{A}$ modulo $\mathscr{I}$.

Let φ be a homomorphism from an algebra $\mathscr{A}$ onto an algebra $\mathscr{B}$. Then, clearly, the *kernel* (*i.e.*, the set of all elements X of $\mathscr{A}$ such that $\varphi(X)$ is the zero element of $\mathscr{B}$) $\mathscr{K}$ of φ is an ideal of $\mathscr{A}$. Moreover, (*see* page 46).

$$\mathscr{A}/\mathscr{K} \cong \mathscr{B}. \tag{79}$$

Furthermore, based on Corollary 5 on page 54 we have

LEMMA 11. *Let $\mathscr{I}$ be an m dimensional ideal of an n dimensional algebra $\mathscr{A}$ over $\mathfrak{F}$. Then $\mathscr{A}/\mathscr{I}$ is an $n - m$ dimensional algebra over $\mathfrak{F}$.*

Also, based on the definition of a supplementary sum (*see* page 47) and Theorem 5 on page 48 we have

LEMMA 12. *Let $\mathscr{I}$ be an ideal and $\mathscr{B}$ a subalgebra of an algebra $\mathscr{A}$ such that*

$$\mathscr{A} = \mathscr{I} \dotplus \mathscr{B}.$$

Then

$$\mathscr{A}/\mathscr{I} \cong \mathscr{B}.$$

The relationship of subalgebras and ideals of a quotient algebra $\mathscr{A}/\mathscr{I}$ to the subalgebras and ideals of $\mathscr{A}$ is given by the following lemma.

LEMMA 13.   *Let $\mathscr{I}$ be an ideal of an algebra $\mathscr{A}$. Then to every sub-algebra $\mathscr{B}$ of $\mathscr{A}$ such that $\mathscr{I} \subset \mathscr{B}$ there corresponds a unique subalgebra $\mathscr{B}/\mathscr{I}$ of $\mathscr{A}/\mathscr{I}$ and vice versa. Moreover, $\mathscr{B}$ is an ideal of $\mathscr{A}$ if and only if $\mathscr{B}/\mathscr{I}$ is an ideal of $\mathscr{A}/\mathscr{I}$ in which case*

$$\mathscr{A}/\mathscr{B} \cong (\mathscr{A}/\mathscr{I})/(\mathscr{B}/\mathscr{I}).$$

The proof of the above lemma is straightforward and is left to the reader.

An ideal $\mathscr{M}$ of an algebra $\mathscr{A}$ is called *maximal* (*see* page 11) if $\mathscr{M} \neq \mathscr{A}$ and $\mathscr{M}$ is contained in no ideal of $\mathscr{A}$ except for $\mathscr{M}$ and $\mathscr{A}$.

LEMMA 14.   *An ideal $\mathscr{M}$ of an algebra $\mathscr{A}$ is a maximal ideal of $\mathscr{A}$ if and only if $\mathscr{A}/\mathscr{M}$ is a simple algebra or a one dimensional zero algebra.*

Proof.   Let $\mathscr{M}$ be maximal and $\mathscr{A}/\mathscr{M}$ not a one dimensional zero algebra. By Lemma 13, the only ideals of $\mathscr{A}/\mathscr{M}$ are $0(\mathscr{M}/\mathscr{M})$ and $\mathscr{A}/\mathscr{M}$. Thus, in view of Definition 3 we see that $\mathscr{A}/\mathscr{M}$ is simple. Next, if $\mathscr{A}/\mathscr{M}$ is one dimensional then by Lemma 11 we see that $\dim \mathscr{A} - \dim \mathscr{M} = 1$ and therefore $\mathscr{M}$ is maximal. On the other hand, if $\mathscr{A}/\mathscr{M}$ is simple, in view of Lemma 13, there is no ideal $\mathscr{R}$ of $\mathscr{A}$ such that $\mathscr{M} \subset \mathscr{R}$ and $\mathscr{M} \neq \mathscr{R}$ and $\mathscr{R} \neq \mathscr{A}$. Thus, $\mathscr{M}$ is maximal as desired.

Exercises

1. Prove Lemma 8 and Corollary 3.

2. Prove (63), (64), (65) and (66).

3. Prove that the centralizer of a nonempty subset of an algebra $\mathscr{A}$ is a sub-albegra of $\mathscr{A}$. Prove also that the center of $\mathscr{A}$ is a commutative subalgebra of $\mathscr{A}$.

4. Prove (69) and (70).

5. Prove (71), (72) and (73).

6. Considering notation (56) prove that $\mathscr{A}^2 = \mathscr{A}$ if $\mathscr{A}$ is a simple algebra.

7. Let $\mathscr{A}$ be a simple algebra. Prove that there exists an element A of $\mathscr{A}$ such that $\mathscr{A}A\mathscr{A} = \mathscr{A}$.

8. Prove that if an algebra $\mathscr{A}$ over a field $\mathfrak{F}$ has a unit element then conditions (73), (74) and (75) are equivalent to the single condition

$$\varphi(pX)(qY+rZ) = pq\varphi(X)\varphi(Y) + pr\varphi(X)\varphi(Z)$$

for every element X, Y, Z of $\mathscr{A}$ and p, q, r of $\mathfrak{F}$.

9. Prove that conditions (73), (74) and (75) are mutually independent.

10. Prove that in view of (76) and (78) the quotient space (77) is an algebra.

11. Prove (79).

12. Prove Lemmas 11, 12 and 13.

13. Prove that any $n-1$ dimensional ideal of an $n > 1$ dimensional algebra $\mathscr{A}$ is a maximal ideal of $\mathscr{A}$.

3.4. Total Matrix Algebras and the Canonical Forms of Matrices

In Section 2.3, we showed that the set $\mathfrak{F}_n$ of all n by n matrices with entries in a field $\mathfrak{F}$ is an n^2 dimensional vector space over $\mathfrak{F}$ with respect to the usual matrix addition and the usual multiplication of a matrix by a scalar (*i.e.*, an element of $\mathfrak{F}$). Also, in Section 1.5, we showed that $\mathfrak{F}_n$ is a ring with respect to the usual matrix addition and multiplication.

Thus, the set $\mathfrak{F}_n$ of all n by n matrices with entries in a field $\mathfrak{F}$ is an n^2 dimensional algebra over $\mathfrak{F}$. The algebra $\mathfrak{F}_n$ plays a fundamental role in the theory of algebras. Because, as shown in the next section, every algebra is isomorphic to a subalgebra of $\mathfrak{F}_n$ for some n and some $\mathfrak{F}$.

For the sake of generality we introduce the following definition.

DEFINITION 4. *An algebra over a field $\mathfrak{F}$ is called a total n by n matrix algebra over $\mathfrak{F}$ if it is isomorphic to the algebra $\mathfrak{F}_n$ of all n by n matrices with entries in $\mathfrak{F}$.*

When no confusion is likely to arise we shall denote a total n by n matrix algebra by $\mathscr{M}_n$ or simply, by $\mathscr{M}$.

REMARK 1. *In view of Definition 4 we see that there is no loss of generality if whenever convenient we treat a total n by n matrix algebra $\mathscr{M}_n$ over a field $\mathfrak{F}_n$ as the algebra $\mathfrak{F}_n$ of all n by n matrices with entries in $\mathfrak{F}$.*

Thus, a total matrix algebra $\mathscr{M}_n$ over a field $\mathfrak{F}$ is of dimension n^2. Moreover, $\mathscr{M}_n$ has a unity. Furthermore, $\mathscr{M}_n$ is a simple algebra (*see* Corollary 1, page 32,) and the center of $\mathscr{M}_n$ is isomorphic to $\mathfrak{F}$ (*see* Corollary 2, page 32). Thus, $\mathscr{M}_n$ is a central simple algebra over $\mathfrak{F}$ (*see* Definition 2, page 83).

Also, as mentioned on page 30, for $n > 1$ a total matrix algebra $\mathscr{M}_n$ has divisors of zero, is not commutative, has nonzero nilpotent elements and has nonzero idempotent elements which are not necessarily equal to the unity of $\mathcal{M}_n$. Clearly, for $n = 1$ a total matrix algebra $\mathcal{M}_n$ over a field $\mathfrak{F}$ is isomorphic to $\mathfrak{F}$.

LEMMA 15.   *An algebra $\mathcal{A}$ over a field $\mathfrak{F}$ is a total matrix algebra $\mathcal{M}_n$ over $\mathfrak{F}$ if and only if $\mathcal{A}$ has a basis consisting of n^2 elements A_{ij} with $i, j = 1, 2, \ldots, n$ such that*

$$A_{ij}A_{jk} = A_{ik} \quad and \quad A_{ij}A_{hk} = 0 \quad if \quad j \neq h.$$

Proof.   In view of Definition 4, the proof of the lemma follows from the fact that the n by n matrices E_{ij} (*see* Remark 2, page 56) form a basis of the n^2 dimensional algebra $\mathfrak{F}_n$ of all n by n matrices over $\mathfrak{F}$.

Let $\mathcal{M}_n$ be a total matrix algebra over a field $\mathfrak{F}$ and let P be a nonsingular element of $\mathcal{M}_n$. As mentioned on page 86 the mapping φ defined by

$$\varphi(M) = PMP^{-1} \tag{80}$$

from $\mathcal{M}_n$ onto $\mathcal{M}_n$ is an automorphism of $\mathcal{M}_n$ which is also called (*see* page 86) an inner automorphism of $\mathcal{M}_n$.

We shall see later that every automorphism of $\mathcal{M}_n$ is an inner automorphism. Therefore, mapping (80) is of particular importance in the study of algebras.

In terms of matrices we interpret mapping (80) as a mapping which transforms an n by n matrix M (with entries in $\mathfrak{F}$) into an n by n matrix PMP^{-1} where P is a nonsingular n by n matrix (with entries in $\mathfrak{F}$). In this connection we introduce the following definition.

DEFINITION 5.   *Let M and A be n by n matrices with entries in a field $\mathfrak{F}$. Then M is called similar in $\mathfrak{F}$ to A if there exists a nonsingular matrix P with entries in $\mathfrak{F}$ such that*

$$PMP^{-1} = A. \tag{81}$$

Clearly, similarity (in $\mathfrak{F}$) is an equivalence relation in the set $\mathfrak{F}_n$ of all n by n matrices over $\mathfrak{F}$. Thus, if (81) holds then we call M and A similar (in $\mathfrak{F}$) matrices and we denote this by

$$M \approx A \qquad (\text{in } \mathfrak{F})$$

or by

$$M \approx A \tag{82}$$

if the underlying field $\mathfrak{F}$ is immaterial for our purposes.

Recalling the definition of the invariant factors of a matrix (*see* pages 34, 35, 36 and 63) we next prove a basic theorem.

THEOREM 3. *Let M be an n by n matrix over a field* $\mathfrak{F}$. *Then the invariant factors of the n by n matrix Ix* $-$ *M are*

$$f_1(x), \quad \underbrace{1, \quad 1, \dots, 1}_{n-1 \text{ ones}} \tag{83}$$

with

$$f_1(x) = x^n + a_{n-1}x^{n-1} + a_{n-2}x^{n-2} + \dots + a_2x^2 + a_1x + a_0 \tag{84}$$

if and only if M is similar (in $\mathfrak{F}$ *) to the matrix A where*

$$A = \begin{pmatrix} 0 & 1 & 0 & \cdots & 0 & 0 \\ 0 & 0 & 1 & \cdots & 0 & 0 \\ \cdot & \cdot & \cdot & \cdots & \cdot & \cdot \\ 0 & 0 & 0 & \cdots & 0 & 1 \\ -a_0 & -a_1 & -a_2 & \cdots & -a_{n-2} & -a_{n-1} \end{pmatrix} \tag{85}$$

Proof. If $M \approx A$ then (81) holds. Comparing (69) on page 63 with (83) and (84) and comparing (68) on page 63 with (85), in view of Lemma 12, page 36 and (81), we see that there exist nonsingular matrices $S(x)$ and $Q(x)$ over $\mathfrak{F}[x]$ such that

$$S(x)(P(Ix-M)P^{-1})Q(x) = S(x)(Ix-A)Q(x) = \text{diag}\{f_1(x), 1, \dots, 1\} \tag{86}$$

where diag $\{f(x), 1, \dots, 1\}$ stands for an n by n diagonal matrix (*see* page 30) whose diagonal (from top to bottom) reads $f_1(x), 1, 1, \dots, 1$. Clearly, $S(x)P$ as well as $P^{-1}Q(x)$ is an n by n nonsingular matrix over $\mathfrak{F}[x]$. Thus, in view of Lemma 12, page 36 we see that the invariant factors of $Ix - M$ are as given in (83).

Now, let the invariant factors of $Ix - M$ be as given in (83). But then, in view of (69) and (68) on page 63, we see that the matrices $Ix - M$ and $Ix - A$ have the same invariant factors. Therefore, by Corollary 5, page 36 there exist nonsingular matrices $S(x)$ and $Q(x)$ such that

$$S(x)(Ix-M)Q(x) = Ix - A. \tag{87}$$

Let us observe that $S(x)$ as well as $Q(x)$ can be regarded as a polynomial (in the indeterminate x) over the ring $\mathfrak{F}_n$ of all n by n matrices. Thus, by (45) and (44) on page 25 we have

$$S(x) = (Ix-A)S_0(x) + P \quad \text{and} \quad Q(x) = Q_0(x)(Ix-A) + Q \tag{88}$$

where $P = S_L(A)$ and $Q = Q_R(A)$ according to (41) and (40) on page 24.

Substituting (88) in (87) we obtain

$$(Ix - A) - (P(Ix - M)Q) = (Ix - A)H(x)(Ix - A) \qquad (89)$$

where

$$H(x) = S_0(x)S^{-1}(x) + Q^{-1}(x)Q_0(x) - S_0(x)(Ix - M)Q_0(x).$$

But then in (89) we must have $H(x) = 0$ since otherwise the right side of the equality sign in (89) would be of degree (in x) at least 2 and the left side of degree at most 1. Consequently, from (89) it follows that

$$P(Ix - M)Q = Ix - A \qquad (90)$$

where M is an n by n matrix over $\mathfrak{F}$ by assumption, A is an n by n matrix over $\mathfrak{F}$ by (85) and P as well as Q is an n by n matrix over $\mathfrak{F}$ by (88). Equating the coefficients of x on both sides of $=$ in (90) we obtain $PQ = I$ implying

$$PMP^{-1} = A.$$

Thus, indeed $M \approx A$ and the theorem is proved.

Let us recall (*see* page 63) that the matrix given by (85) is called the *companion matrix* of the polynomial $f_1(x)$ given by (84).

COROLLARY 4. *Let $f_1(x)$ be a monic polynomial of degree n with coefficients in a field $\mathfrak{F}$. Then an n by n matrix M over $\mathfrak{F}$ is similar (in $\mathfrak{F}$) to the companion matrix of $f_1(x)$ if and only if the invariant factors of $Ix - M$ are*

$$f_1(x), \quad \underbrace{1, 1, \ldots, 1}_{n-1 \text{ ones}}$$

A generalization of Theorem 3 and Corollary 4 is given below.

THEOREM 4. *Let $f_1(x), f_2(x), \ldots, f_r(x)$ be monic polynomials with coefficients in a field $\mathfrak{F}$ and of positive degrees $n_1, n_2, \ldots, n_r$ respectively. Moreover, let f_{i+1} divide f_i for $1 \leqslant i \leqslant r - 1$ and let $n = n_1 + n_2 + \cdots + n_r$. Then*

$$f_1(x), \quad f_2(x), \ldots, f_r(x), \quad \underbrace{1, 1, \ldots, 1}_{n-r \text{ ones}} \qquad (91)$$

are the invariant factors of an n by n matrix $Ix - M$ if and only if M is similar (in $\mathfrak{F}$) to the matrix

$$C = \text{diag}\{C_1, C_2, \ldots, C_r\} \qquad (92)$$

where C is an n by n matrix which is broken up into blocks whose non-diagonal blocks are zero matrices and whose diagonal blocks are matrices

$C_1, C_2, \ldots, C_r$ *where for* $i = 1, 2, \ldots, r$

$$C_i = \text{companion matrix of } f_i(x).$$

Proof. Let us remark that by means of operations of type (i), (ii) and (iii) described on page 33, it is easy to verify that the matrix $Ix - C$ (where C is given as in (92)) is rationally equivalent (in $\mathfrak{F}[x]$) to a diagonal matrix whose diagonal is given by (91). Thus, the invariant factors of the matrix $Ix - C$ are as given in (91). But then, in view of this remark, it is readily seen that the proof of Theorem 4 can be given similar to that of Theorem 3.

DEFINITION 6. *Let* $f_1(x), f_2(x), \ldots, f_r(x)$ *be monic polynomials as described in Theorem 4. Then the matrix* C *as given by* (92) *is called the rational canonical matrix corresponding to* $f_1(x), f_2(x), \ldots, f_r(x)$. *Moreover, the matrix* C *is called the rational canonical form of any matrix* M *the invariant factors of whose characteristic matrix (i.e., the matrix* $Ix - M$) *are as given in* (91).

Let us give an example. As shown on page 63 the invariant factors of the characteristic matrix $Ix - M$ of the matrix M where

$$M = \begin{pmatrix} 0 & 2 & -1 \\ 1 & -1 & 1 \\ 1 & -2 & 2 \end{pmatrix}$$

are: $x^2 - 1$, $x - 1$, 1. Therefore (say over the field of reals) the rational canonical form of M is the matrix C where

$$C = \left(\begin{array}{cc|c} 0 & 1 & 0 \\ 1 & 0 & 0 \\ \hline 0 & 0 & 1 \end{array} \right)$$

Clearly, according to Theorem 4 every matrix is similar to its unique rational canonical form. Thus, in particular

$$\begin{pmatrix} 0 & 1 & -1 \\ 1 & -1 & 1 \\ 1 & -2 & 2 \end{pmatrix} \approx \left(\begin{array}{cc|c} 0 & 1 & 0 \\ 1 & 0 & 0 \\ \hline 0 & 0 & 1 \end{array} \right)$$

In view of Theorem 4 and Definition 6 we have the obvious corollary below.

COROLLARY 5. *Let* A *and* B *be* n *by* n *matrices with entries in a field* $\mathfrak{F}$. *Then the following statements are equivalent.*
(i) A *is similar to* B (*in* $\mathfrak{F}$).

(ii) *The characteristic matrices* $(Ix - A)$ *and* $(Ix - B)$ *of the matrices A and B have the same invariant factors (in* $\mathfrak{F}[x]$*).*

(iii) *A and B have the same rational canonical form (in* $\mathfrak{F}$*).*

Next, we introduce the following definition

DEFINITION 7. *Let $p^n(x)$ be the n-th power $(n = 1, 2, \ldots)$ of an irreducible monic polynomial $p(x)$ with coefficients in a field $\mathfrak{F}$. Then the matrix $\mathfrak{F}$ given by*

$$J = \begin{pmatrix} \boxed{C_1} & 1 & & & \\ & \boxed{C_2} & 1 & & \Large 0 \\ & & & \ddots & \\ & & & & 1 \\ \Large 0 & & & & \boxed{C_n} \end{pmatrix} \tag{93}$$

where for $i = 1, 2, \ldots, n$

$$C_i = \text{the companion matrix of } p(x)$$

is called the Jordan companion matrix (in $\mathfrak{F}$) of $p^n(x)$.

For instance, over the field of reals the Jordan companion matrix of $(x^2 + 2x + 3)^2$ is

$$\begin{pmatrix} 0 & 1 & 0 & 0 \\ -3 & -2 & 1 & 0 \\ \hline 0 & 0 & 0 & 1 \\ 0 & 0 & -3 & -2 \end{pmatrix} \tag{94}$$

LEMMA 16. *Let $p^n(x)$ be the n-th power $(n = 1, 2, \ldots)$ of an irreducible monic polynomial $p(x)$ with coefficients in a field $\mathfrak{F}$. Then the Jordan companion matrix J of $p^n(x)$ is similar (in $\mathfrak{F}$) to the companion matrix C of $p^n(x)$.*

Proof. By means of operations of type (i), (ii) and (iii) described on page 33, it is easy to verify that the invariant factors of the matrix $Ix - J$ where J is given by (93) is

$$P^n(x), \quad \underbrace{1, 1, \ldots, 1}_{k-1 \text{ ones}}$$

where $k = $ degree of $p^n(x)$. Thus, in view of Corollary 5 we see that $J \approx C$, as desired.

For instance, since over the field of reals

$$(x^2 + 2x + 3)^2 = x^4 + 4x^3 + 10x^2 + 12x + 9$$

in view of (94) we have

$$
\begin{pmatrix}
0 & 1 & 0 & 0 \\
-3 & -2 & 1 & 0 \\
0 & 0 & 0 & 1 \\
0 & 0 & -3 & -2
\end{pmatrix}
\approx
\begin{pmatrix}
0 & 1 & 0 & 0 \\
0 & 0 & 1 & 0 \\
0 & 0 & 0 & 1 \\
-9 & -12 & -10 & -4
\end{pmatrix}
$$

Recalling the definition of the elementary divisors of a matrix (*see* pages 34, 35, 36 and 63) we prove the following theorem.

THEOREM 5. *Let* $p_1^{k_1}(x)$, $p_2^{k_2}(x), \ldots, p_m^{k_m}(x)$ *be the nontrivial elementary divisors of an n by n matrix* $Ix - M$ *where M is an n by n matrix over a field* $\mathfrak{F}$. *Then the matrix M is similar (in* $\mathfrak{F}$*) to the matrix*

$$J = \text{drag}\{J_1, J_2, \ldots, J_m\} \tag{95}$$

where for $i = 1, 2, \ldots, m$

$$J_i = Jordan\ companion\ matrix\ of\ p_i^{k_m}(x).$$

Proof. As mentioned on page 35, the elementary divisors of a matrix determine the invariant factors of that matrix and vice versa. Thus, the nontrivial elementary divisors $p_i^{k_m}(x)$ of the matrix $Ix - M$ determine the invariant factors $f_1(x), f_2(x), \ldots, f_r(x)$ of $Ix - M$. On the other hand, by means of operations of type (i), (ii) and (iii) described on page 00, it is easy to verify that the nontrivial invariant factors of the matrix $(Ix - J)$ are $f_1(x), f_2(x), \ldots, f_r(x)$. Thus, by Corollary 5 we see that $J \approx M$, as desired.

DEFINITION 8. *Let*

$$p_1^{k_1}(x), p_2^{k_2}(x), \ldots, p_m^{k_m}(x) \tag{96}$$

be respectively the $k_1, k_2, \ldots, k_m$*-th powers of the (not necessarily distinct) irreducible monic polynomials* $p_1(x)$, $p_2(x), \ldots, p_m(x)$ *with coefficients in a field* $\mathfrak{F}$. *Then the matrix J as given by (95) is called the Jordan canonical matrix corresponding to the polynomials listed in (96). Moreover, the matrix J is called the Jordan canonical form of any matrix M the nontrivial elementary divisors (in* $\mathfrak{F}$*) of whose characteristic matrix (i.e., the matrix* $Ix - M$*) are the polynomials listed in (96).*

In view of Theorem 5, Corollary 5 and Definition 8 we have:

COROLLARY 6. *Let A and B be n by n matrices with entries in a field*
$\mathfrak{F}$. *Then the following statements are equivalent.*
  (i) *A is similar to B (in* $\mathfrak{F}$).
  (ii) *The characteristic matrices* $(Ix-A)$ *and* $(Ix-B)$ *of the matrices A
and B have the same elementary divisors (in* $\mathfrak{F}[x]$).
  (iii) *A and B have the same Jordan canonical form (in* $\mathfrak{F}$).

Let us give some examples. Consider the matrix M given on page 93.
The invariant factors of $Ix-M$ are: x^2-1, $x-1$, 1. Therefore, the
elementary divisors (over any field) of $Ix-M$ are

$$x+1, \quad x-1, \quad x-1, \quad 1.$$

Consequently,

$$\begin{pmatrix} 0 & 2 & -1 \\ 1 & -1 & 1 \\ 1 & -2 & 2 \end{pmatrix} \approx \begin{pmatrix} 0 & 1 & 0 \\ 1 & 0 & 0 \\ 0 & 0 & 1 \end{pmatrix} \approx \begin{pmatrix} -1 & 0 & 0 \\ 0 & 1 & 0 \\ 0 & 0 & 1 \end{pmatrix}$$

where in the above the second and the third matrices are respectively the
rational canonical and the Jordan canonical form of the first matrix.
  Again, let G be a 10 by 10 matrix over the reals such that the elementary divisors of $Ix-G$ are:

$$(x^2+2x+3)^2, \quad (x^2+1)^2, \quad (x^2+1), \quad 1,1,1,1,1,1,1,1.$$

Then clearly, the invariant factors of $Ix-G$ are:

$$(x+2x+3)^2(x+1)^2, \quad (x^2+1), \quad 1,1,1,1,1,1,1,1.$$

Observing that

$$(x^2+2x+3)^2(x^2+1)^2$$
$$= x^8+4x^7+12x^6+20x^5+40x^4+28x^3+48x^2+12x+19$$

we obtain the rational canonical form C and the Jordan canonical form J
of the matrix G as follows

$$C = \left(\begin{array}{cccccccc|cc} 0 & 1 & 0 & 0 & 0 & 0 & 0 & 0 & 0 & 0 \\ 0 & 0 & 1 & 0 & 0 & 0 & 0 & 0 & 0 & 0 \\ 0 & 0 & 0 & 1 & 0 & 0 & 0 & 0 & 0 & 0 \\ 0 & 0 & 0 & 0 & 1 & 0 & 0 & 0 & 0 & 0 \\ 0 & 0 & 0 & 0 & 0 & 1 & 0 & 0 & 0 & 0 \\ 0 & 0 & 0 & 0 & 0 & 0 & 1 & 0 & 0 & 0 \\ 0 & 0 & 0 & 0 & 0 & 0 & 0 & 1 & 0 & 0 \\ -19 & -12 & -48 & -28 & -40 & -20 & -12 & -4 & 0 & 0 \\ \hline 0 & 0 & 0 & 0 & 0 & 0 & 0 & 0 & 0 & 1 \\ 0 & 0 & 0 & 0 & 0 & 0 & 0 & 0 & -1 & 0 \end{array}\right)$$

$$J = \begin{pmatrix} 0 & 1 & 0 & 0 & 0 & 0 & 0 & 0 & 0 & 0 \\ -3 & -2 & 1 & 0 & 0 & 0 & 0 & 0 & 0 & 0 \\ 0 & 0 & 0 & 1 & 0 & 0 & 0 & 0 & 0 & 0 \\ 0 & 0 & -3 & -2 & 0 & 0 & 0 & 0 & 0 & 0 \\ 0 & 0 & 0 & 0 & 0 & 1 & 0 & 0 & 0 & 0 \\ 0 & 0 & 0 & 0 & -1 & 0 & 1 & 0 & 0 & 0 \\ 0 & 0 & 0 & 0 & 0 & 0 & 0 & 1 & 0 & 0 \\ 0 & 0 & 0 & 0 & 0 & 0 & -1 & 0 & 0 & 0 \\ 0 & 0 & 0 & 0 & 0 & 0 & 0 & 0 & 0 & 1 \\ 0 & 0 & 0 & 0 & 0 & 0 & 0 & 0 & -1 & 0 \end{pmatrix}$$

Obviously, $G \approx C \approx J$.

Recalling the definition of the *rank* (*i.e.*, the number of the linearly independent rows) of a matrix with entries in a field, we prove the following theorem.

LEMMA 17.   *Let E be an n by n matrix with entries in a field $\mathfrak{F}$. Then E is an idempotent matrix of rank m if and only if E is similar (in $\mathfrak{F}$) to the diagonal matrix*

$$U = \begin{pmatrix} 1 & & & & & & \\ & 1 & & & & \mathbf{0} & \\ & & \ddots & & & & \\ & & & 1 & & & \\ & & & & 0 & & \\ & \mathbf{0} & & & & \ddots & \\ & & & & & & 0 \end{pmatrix} \begin{matrix} \\ \\ \\ m\text{-th row} \\ \\ \\ n\text{-th row} \end{matrix} \qquad (97)$$

which is the Jordan canonical form of E.

Proof.   Let E be an n by n idempotent matrix of rank m. By Corollary 1, page 74, the minimum equation of E is $X = 0$, or $X = 1$, or $X^2 = X$. Thus, the first invariant factor (Definition 5, page 62) of $Ix - E$ is x, or $x - 1$, or $x^2 - x$. Consequently, the nontrivial elementary divisors of $Ix - E$ are

$$\underbrace{x - 1, \quad x - 1, \ldots, x - 1.}_{m} \quad \underbrace{x, \quad x, \ldots, x}_{n - m}$$

and by Theorem 5, the Jordan canonical form of E (to which it is similar) is matrix U, as given in (97).

Conversely, since similarity preserves idempotence and the rank of a matrix and since U is an idempotent matrix of rank m, we see that if $E \approx U$ then E is an idempotent matrix of rank m.

In particular, if in the above lemma $n = m$ then $U = I$ (*i.e.*, the n by n unit matrix) and if $m = 0$ then $U = 0$ (*i.e.*, the n by n zero matrix).

LEMMA 18.   *Let N be an n by n matrix with entries in a field $\mathfrak{F}$. Then N is a nilpotent matrix of index $k > 0$ if and only if N is similar (in $\mathfrak{F}$) to the n by n matrix*

$$V = \text{diag}\,\{V_1, V_2, \ldots, V_h\} \tag{98}$$

where

$$V_i = \begin{pmatrix} 0 & 1 & 0 & \cdots & 0 \\ 0 & 0 & 1 & \cdots & 0 \\ \cdot & \cdot & \cdot & \cdots & \cdot \\ 0 & 0 & 0 & \cdots & 1 \\ 0 & 0 & 0 & \cdots & 0 \end{pmatrix}$$

such that V_1 is a k by k matrix and for $i > 1$ the matrices V_i are m_i by m_i with $m_i \leqslant k$.

Proof.   Let N be an n by n nilpotent matrix of index $k > 0$. By Lemma 4, page 74, the first invariant factor (Definition 5, page 62) of $Ix - N$ is X^k. Consequently, the nontrivial invariant factors of $Ix - N$ are

$$x^k, \quad x^{m_2}, \ldots, x^{m_h}$$

for some natural number h such that $m_i \leqslant k$ for $i > 1$. But then by Theorem 4, the rational canonical form of N (to which it is similar) is matrix V, as given in (98).

Conversely, since similarity preserves nilpotence and the index of nilpotence of a matrix and since V is a nilpotent matrix of index k (by virtue of the fact that V_1 is a k by k matrix and $m_i \leqslant k$ for $i > 1$) we see that if $N \approx V$ then N is a nilpotent matrix of index k.

In particular, if in the above lemma, $k = 1$ then $V = 0$.

THEOREM 6.   *Let E and H be n by n orthogonal (i.e., $EH = 0$) idempotent matrices respectively of rank e and h over a field $\mathfrak{F}$. Then there* *exists a nonsingular matrix P over $\mathfrak{F}$ such that*

$$PEP^{-1} = \text{diag}\,\{I_e, 0_{n-e}\} = \begin{pmatrix} 1 & & & & & \\ & \ddots & & & & \\ & & 1 & & & 0 \\ & & & 0 & & \\ & & & & \ddots & \\ & 0 & & & & \ddots \\ & & & & & & 0 \end{pmatrix}$$

and $\hspace{12cm}$ (99)

$$PHP^{-1} = \text{diag}\,\{0_e, I_h, 0_{n-e-h}\} = \begin{pmatrix} 0 & & & & & \\ & \ddots & & & & \\ & & 0 & & & 0 \\ & & & 1 & & \\ & & & & \ddots & \\ & 0 & & & & 1 \\ & & & & & & 0 \\ & & & & & & & \ddots \\ & & & & & & & & 0 \end{pmatrix}$$

$\hspace{13cm}$ (100)

where I_j is a j by j unit matrix and 0_j is a j by j zero matrix.
   Moreover, $E + H$ is an idempotent matrix of rank $e + h \leq n$.

   Proof. By Lemma 17, there exists a nonsingular matrix Q such that (using notation similar to that introduced in (92), page 92)

$$QEQ^{-1} = \text{diag}\,\{I_e, 0_{n-e}\}. \hspace{3cm} (101)$$

Consider QHQ^{-1}. Since $EH = 0$, we have

$$(QEQ^{-1})(QHQ^{-1}) = QEHQ^{-1} = 0.$$

From the above, in view of (101), it follows that

$$QHQ^{-1} = \text{diag}\,\{0_e, M\}$$

where M is an $n-e$ by $n-e$ idempotent matrix of rank h (since H is an idempotent matrix of rank h).

But then by Lemma 17, there exists a nonsingular $n-e$ by $n-e$ matrix S such that

$$SMS^{-1} = \text{diag}\,\{I_h, 0_{n-e-h}\}. \qquad (102)$$

Let us define

$$P = \text{diag}\,\{Q, S\}.$$

Then, in view of (101) and (102), it can be readily verified that PEP^{-1} and PHP^{-1} are respectively equal to the matrices given in (99) and (100).

The idempotence of $E+H$ and the fact that the rank of $E+H$ is $e+h$ follows easily from (99) and (100).

Theorem 6 yields the following corollary.

COROLLARY 7. *Let $E_1, E_2, \ldots, E_m$ be n by n pairwise orthogonal idempotent matrices respectively of rank $r_1, r_2, \ldots, r_m$ over a field $\mathfrak{F}$. Then there exists a nonsingular matrix P over $\mathfrak{F}$ such that*

$$PE_iP^{-1} = \text{diag}\,\{0_u, I_{r_i}, 0_v\}$$

where $u = r_1 + r_2 + \cdots r_{i-1}$ and $v = r_{i+1} + r_{i+2} + \cdots + r_m$.

Recalling the definition of an automorphism of an algebra (*see* page 86) we prove the following theorem.

THEOREM 7. *Every automorphism of a total matrix algebra is an inner automorphism.*

Proof. In view of Remark 1, page 89, we prove the theorem for the algebra $\mathfrak{F}_n$ of all n by n matrices with entries in a field $\mathfrak{F}$.

Let φ be an automorphism of $\mathfrak{F}_n$. Clearly, the n by n matrices E_{ij} form a basis of $\mathfrak{F}_n$ (*see* Remark 2, page 56). Thus, to prove the theorem it is enough to show that there exists a nonsingular n by n matrix P over $\mathfrak{F}$ such that

$$\varphi(E_{ij}) = PE_{ij}P^{-1} \qquad (i,j = 1, 2, \ldots, n). \qquad (103)$$

However, $\varphi(E_{ii})$ for $i = 1, 2, \ldots, n$ are pairwise orthogonal idempotent matrices each of rank 1. Therefore, by Corollary 7, there exists a nonsingular matrix Q over $\mathfrak{F}$ such that

$$Q\varphi(E_{ii})Q^{-1} = E_{ii} \qquad (i = 1, 2, \ldots, n). \qquad (104)$$

On the other hand, we have

$$E_{1j} = E_{11}E_{1j}E_{jj} \qquad (j = 1, 2, \ldots n).$$

and since φ is an automorphism, in view of (104), we obtain

$$Q\varphi(E_{1j})Q^{-1} = Q\varphi(E_{11}E_{1j}E_{jj})Q^{-1} = E_{11}Q\varphi(E_{1j})Q^{-1}E_{jj}.$$

The above shows that

$$Q\varphi(E_{1j})Q^{-1} = q_{1j}E_{1j}$$

for some nonzero element q_{ij} of $\mathfrak{F}$. Clearly, $q_{11} = 1$.

Consider the diagonal matrix

$$S = \operatorname{diag}\{1, q_{12}, \ldots, q_{1n}\}.$$

Then it is easy to verify that with $P = Q^{-1}S^{-1}$ equality (103) holds.

Exercises

1. Prove that for every natural number m with $1 \leqslant m \leqslant n$ a total matrix algebra $\mathcal{M}_n$ has subalgebras isomorphic to $\mathcal{M}_m$.

2. Let M be an n by n matrix over $\mathfrak{F}$ and I the unit n by n matrix. Using notation (82), prove that $M \approx I$ (in $\mathfrak{F}$) if and only if $M = I$.

3. Let A be an n by n matrix over the field of rational numbers such that the nontrivial invariant factors of $Ix - A$ are $x^2 - 5x + 6$ and $x - 2$. Determine n and the number of the trivial invariant factors of $Ix - A$. Also, determine the elementary divisors of $Ix - A$.

4. Write the rational as well as the Jordan canonical form of matrix A in Problem 3.

5. Let B be an n by n matrix over the field of rationals and let the nontrivial elementary divisors of $Ix - B$ be $(x^2 + 1)^2$, $(X + 1)$ and x. Determine n and the number of the trivial elementary divisors of $Ix - B$. Also, determine the invariant factors of B and write the rational as well as the Jordan canonical form of matrix B.

6. Let M be an n by n matrix over a field $\mathfrak{F}$ such that the characteristic polynomial det $(Ix - M)$ of M has n distinct roots $r_1, r_2, \ldots, r_n$ (called the *characteristic roots* of M). Using notation (86), prove that M is similar (in $\mathfrak{F}$) to diag $\{r_1, r_2, \ldots, r_n\}$.

7. Give an example of a 3 by 3 matrix M over the field of rationals $\mathfrak{F}$ such that the characteristic roots of M are 2, 2 and 5 and such that M is not similar (in $\mathfrak{F}$) to any diagonal matrix.

8. Let M be a 6 by 6 matrix over a field $\mathfrak{F}$ and let the nontrivial invariant factors of $Ix - M$ be

$$(x^2 - 2)^2 \qquad \text{and} \qquad (x^2 - 2).$$

Determine the Jordan canonical form of M if $\mathfrak{F}$ is the field of rationals. Also, determine the Jordan canonical form of M if $\mathfrak{F}$ is the field of reals.

9. Let M be a 4 by 4 matrix over a field $\mathfrak{F}$ and let the only nontrivial invariant factor of $Ix - M$ be $(x^2 + 1)(x^2 + 2)$. Prove that the rational canonical form of M is the same if $\mathfrak{F}$ is the field of rationals or reals or complex numbers. Also, prove that the Jordan canonical forms of M are pairwise distinct for the three cases mentioned. Moreover, verify that the rational canonical form of M is the same as the Jordan canonical form of M if $\mathfrak{F}$ is the field of rational numbers.

10. Let M be a square matrix over a field $\mathfrak{F}$. Prove that the rational canonical form of M is the same as the Jordan canonical form of M if and only if every nontrivial invariant factor of $Ix - M$ is an irreducible polynomial over $\mathfrak{F}$.

3.5. Matrix Representation of Algebras

In this section we shall encounter another instance of the use of matrices in algebraic theories. In Section 2.3 we showed that the set $\mathfrak{F}_{mn}$ of all m by n matrices over a field $\mathfrak{F}$ can be used to represent any vector space of dimension $\leq mn$ over $\mathfrak{F}$. In Section 3.4 we show that the set $\mathfrak{F}_n$ of all n by n matrices are used to represent the total n by n matrix algebra $\mathfrak{F}_n$ over $\mathfrak{F}$. In a sense, $\mathfrak{F}_n$ is a rather significant and general example of an algebra since, for $n \geq 2$, it is noncommutative, has nontrivial idempotent and nilpotent elements and has nontrivial nonsingular and singular elements. The generality of a total matrix algebra and the use of matrices is emphasized once more by the fact (as mentioned on page 89) that, as shown by the theorem below, every algebra is isomorphic to a subalgebra of a total matrix algebra $\mathfrak{F}_n$ (or $\mathcal{M}_n$) for some n and some $\mathfrak{F}$. In view of this we see that (for all algebraic purposes) in constructing an algebra and studying an algebra we may restrict ourselves to considering a set of square matrices with the usual matrix addition and multiplication.

THEOREM 8. *Let $\mathcal{A}$ be an n dimensional algebra over a field $\mathfrak{F}$. If $\mathcal{A}$ has a unity then $\mathcal{A}$ is isomorphic to a subalgebra of the total n by n matrix algebra $\mathfrak{F}_n$. If $\mathcal{A}$ has no unity then $\mathcal{A}$ is isomorphic to a subalgebra of the total $n+1$ by $n+1$ matrix algebra $\mathfrak{F}_{n+1}$.*

Proof. We prove the theorem for the case $n = 3$ which illustrates the proof in general. Thus, we assume that $\mathcal{A}$ is a three dimensional algebra over a field $\mathfrak{F}$ and that $\{I, A, B\}$ is a basis of $\mathcal{A}$ where I is the unity of $\mathcal{A}$. Let the multiplication table of $\mathcal{A}$ with respect to the basis $\{I, A, B\}$ be given by

	I	A	B
I	$1I + 0A + 0B$	$0I + 1A + 0B$	$0I + 0A + 1B$
A	$0I + 1A + 0B$	$aI + bA + cB$	$gI + hA + kB$
B	$0I + 0A + 1B$	$qI + rA + tB$	$uI + vA + wB$

$$(105)$$

Clearly, we may represent I by $(1, 0, 0)$, A by $(0, 1, 0)$ and B by $(0, 0, 1)$ and rewrite table (105) as follows

	I	A	B
$(1, 0, 0)$	$(1, 0, 0)$	$(0, 1, 0)$	$(0, 0, 1)$
$(0, 1, 0)$	$(0, 1, 0)$	(a, b, c)	(g, h, k)
$(0, 0, 1)$	$(0, 0, 1)$	(q, r, t)	(u, v, w)

$$(106)$$

From table (106) we see that

$$(1, 0, 0) \cdot I = (1, 0, 0); \quad (0, 1, 0) \cdot I = (0, 1, 0); \quad (0, 0, 1) \cdot I = (0, 0, 1).$$

The above equalities suggest substituting

$$for \quad I \quad the\,matrix \quad \begin{pmatrix} 1 & 0 & 0 \\ 0 & 1 & 0 \\ 0 & 0 & 1 \end{pmatrix}. \tag{107}$$

Again, from table (106) we see that

$$(1, 0, 0) \cdot A = (0, 1, 0); \quad (0, 1, 0) \cdot A = (a, b, c); \quad (0, 0, 1) \cdot A = (q, r, t).$$

The above equalities suggest substituting

$$for \quad A \quad the\,matrix \quad \begin{pmatrix} 0 & 1 & 0 \\ a & b & c \\ q & r & t \end{pmatrix}. \tag{108}$$

Finally, from table (106) we see that

$$(1, 0, 0) \cdot B = (0, 0, 1); \quad (0, 1, 0) \cdot B = (g, h, k); \quad (0, 0, 1) \cdot B = (u, v, w).$$

The above equalities suggest substituting

$$for \quad B \quad the\,matrix \quad \begin{pmatrix} 0 & 0 & 1 \\ g & h & k \\ u & v & w \end{pmatrix}. \tag{109}$$

Motivated by (107), (108) and (109), we consider the mapping φ from

$\mathscr{A}$ into the total 3 by 3 matrix algebra $\mathfrak{F}_3$ where for every element S of $\mathscr{A}$ with

$$S = s_1 I + s_2 A + s_3 B \tag{110}$$

we define

$$\varphi(S) = s_1 \varphi(I) + s_2 \varphi(A) + s_3 \varphi(B) \tag{111}$$

where

$$\varphi(I) = \begin{pmatrix} 1 & 0 & 0 \\ 0 & 1 & 0 \\ 0 & 0 & 1 \end{pmatrix} \qquad \varphi(A) = \begin{pmatrix} 0 & 1 & 0 \\ a & b & c \\ q & r & t \end{pmatrix} \qquad \varphi(B) = \begin{pmatrix} 0 & 0 & 1 \\ g & h & k \\ u & v & w \end{pmatrix}. \tag{112}$$

We prove that φ is an algebra isomorphism from $\mathscr{A}$ onto a subalgebra of $\mathfrak{F}_3$.

For every element X of $\mathscr{A}$ with $X = x_1 I + x_2 A + x_3 B$, we let X represent the 1 by 3 matrix (x_1, x_2, x_3), *i.e.*,

$$\overline{X} = (x_1, x_2, x_3). \tag{113}$$

Then from (107) through (113) it follows that for every element S of $\mathscr{A}$

$$\varphi(S) = \mathbf{M} \quad \textit{if and only if} \quad \overline{XS} = \overline{X}\mathbf{M} \quad \textit{for every element } X \textit{ of } \mathscr{A} \tag{114}$$

where, naturally, $\varphi(S)$ as well as $\mathbf{M}$ is a 3 by 3 matrix.

Now, let S and Q be elements of $\mathscr{A}$. But then in view of (113), (114) and the distributivity of multiplication with respect to addition in $\mathscr{A}$ as well as in $\mathfrak{F}_3$ we have:

$$\overline{X(S+Q)} = \overline{XS} + \overline{SQ} = \overline{X}(\varphi(S) + \varphi(Q))$$

which in view of (114) implies

$$\varphi(S+Q) = \varphi(S) + \varphi(Q). \tag{115}$$

Again, in view of (113) and (114) and the associativity of multiplication in $\mathscr{A}$ as well as in $\mathfrak{F}_3$ we have:

$$\overline{X(SQ)} = (\overline{XS})\overline{Q} = (\overline{X}\varphi(S))\varphi(Q) = \overline{X}(\varphi(S)\varphi(Q))$$

which, in view of (114) implies

$$\varphi(SQ) = \varphi(S)\varphi(Q). \tag{116}$$

Finally, for every scalar s we have:

$$\overline{X(sS)} = \overline{X}(s\overline{S}) = \overline{X}(s\varphi(S))$$

which in view of (114) implies

$$\varphi(sS) = s\varphi(S). \tag{117}$$

Comparing (115), (116), (117) with (73), (74), (75) on page 86 we conclude that φ is an algebra homomorphism from $\mathscr{A}$ into $\mathfrak{F}_3$. On the other hand, inspecting the first rows of the matrices given in (107), (108) and (109), we see that these matrices are linearly independent and therefore φ is an algebra isomorphism from $\mathscr{A}$ onto a subalgebra of $\mathfrak{F}_3$, as desired.

Next, let $\mathscr{A}$ be a three dimensional albegra without unity over a field $\mathfrak{F}$ and let $\{G, H, R\}$ be a basis of $\mathscr{A}$. Let the multiplication table of $\mathscr{A}$ with respect to the basis $\{G, H, R\}$ be given by

	G	H	R
G	(g_{11}, g_{12}, g_{13})	(h_{11}, h_{12}, h_{13})	(r_{11}, r_{12}, r_{13})
H	(g_{21}, g_{22}, g_{23})	(h_{21}, h_{22}, h_{23})	(r_{21}, r_{22}, r_{23})
R	(g_{31}, g_{32}, g_{33})	(h_{31}, h_{32}, h_{33})	(r_{31}, r_{32}, r_{33})

(118)

Consider the mapping Φ from $\mathscr{A}$ into the total 4 by 4 matrix algebra $\mathfrak{F}_4$ where for an element X of $\mathscr{A}$ with

$$X = s_1 G + s_2 H + s_3 R$$

we define

$$\Phi(X) = s_1 \Phi(G) + s_2 \Phi(H) + s_3 \Phi(R) \tag{119}$$

and where

$$\Phi(G) = \begin{pmatrix} 0 & 1 & 0 & 0 \\ 0 & g_{11} & g_{12} & g_{13} \\ 0 & g_{21} & g_{22} & g_{23} \\ 0 & g_{31} & g_{32} & g_{33} \end{pmatrix} \quad \Phi(H) = \begin{pmatrix} 0 & 0 & 1 & 0 \\ 0 & h_{11} & h_{12} & h_{13} \\ 0 & h_{21} & h_{22} & h_{23} \\ 0 & h_{31} & h_{32} & h_{33} \end{pmatrix}$$

$$\Phi(R) = \begin{pmatrix} 0 & 0 & 0 & 1 \\ 0 & r_{11} & r_{12} & r_{13} \\ 0 & r_{21} & r_{22} & r_{23} \\ 0 & r_{31} & r_{32} & r_{33} \end{pmatrix} \tag{120}$$

Again, it is easy to verify that the mapping Φ given by (119) and (120) is an algebra isomorphism from $\mathscr{A}$ onto a subalgebra of $\mathfrak{F}_4$, as desired.

Let $\mathscr{S}$ be the three dimensional subalgebra of $\mathfrak{F}_4$ whose basis is $\{\Phi(G), \Phi(H), \Phi(R)\}$ and let $\mathscr{S}_1$ be the four dimensional subalgebra of $\mathfrak{F}_4$ whose basis is $\{I_4, \Phi(G), \Phi(H), \Phi(R)\}$ where I_4 is the unit 4 by 4 matrix. Clearly, $\mathscr{S}_1$ has a unity and from (120) it follows that $\mathscr{S}$ is a subalgebra of $\mathscr{S}_1$. Thus, in effect, we have proved:

COROLLARY 8. *Every n dimensional algebra without a unity is a sub-algebra of an n + 1 dimensional algebra with a unity.*

REMARK 2. *The reader is advised to observe the simple method of constructing the matrices given in* (107), (108), (109) *from table* (105), *and, the matrices given in* (120) *from table* (118).

DEFINITION 9. *The representation of an n dimensional algebra $\mathscr{A}$ with a unity by means of n by n matrices as given by* (111) *and* (112) *is called the regular representation of $\mathscr{A}$ with respect to the basis of $\mathscr{A}$ which is under consideration. Similarly, the representation of an n dimensional algebra $\mathscr{A}$ without unity by means of $n + 1$ by $n + 1$ matrices as given by* (119) *and* (120) *is called the regular representation of $\mathscr{A}$ with respect to the basis of $\mathscr{A}$ which is under consideration.*

For an m by n matrix $\mathbf{M}$ we denote by $\mathbf{M}'$ the *transpose* of $\mathbf{M}$, *i.e.*, $\mathbf{M}'$ is an n by m matrix whose i-th row is the i-th column of $\mathbf{M}$ for $i = 1, 2, \ldots, n$.

Let us consider again table (105). Clearly, we may represent I by $(1, 0, 0)'$, A by $(0, 1, 0)'$ and B by $(0, 0, 1)'$ and rewrite table (105) as follows:

	$(1, 0, 0)'$	$(0, 1, 0)'$	$(0, 0, 1)'$
I	$(1, 0, 0)'$	$(0, 1, 0)'$	$(0, 0, 1)'$
A	$(0, 1, 0)'$	$(a, b, c)'$	$(g, h, k)'$
B	$(0, 0, 1)'$	$(q, r, t)'$	$(u, v, w)'$

From this table we see that

$$I \cdot (1, 0, 0)' = (1, 0, 0)'; \quad I \cdot (0, 1, 0)' = (0, 1, 0)'; \quad I \cdot (0, 0, 1)' = (0, 0, 1)'.$$

The above equalities suggest substituting

$$\text{\textit{for} \quad I \quad \textit{the matrix}} \quad \begin{pmatrix} 1 & 0 & 0 \\ 0 & 1 & 0 \\ 0 & 0 & 1 \end{pmatrix}'. \tag{121}$$

Again, from the above table we see that

$$A \cdot (1, 0, 0)' = (0, 1, 0)';$$

$$A \cdot (0, 1, 0)' = (a, b, c)';$$

$$A \cdot (0, 0, 1)' = (g, h, k)'.$$

The above equalities suggest substituting

$$for \quad A \quad the\ matrix \quad \begin{pmatrix} 0 & 1 & 0 \\ a & b & c \\ g & h & k \end{pmatrix}'. \qquad (122)$$

Finally, from the above table we see that

$$B \cdot (1, 0, 0)' = (0, 0, 1)';$$

$$B \cdot (0, 1, 0)' = (q, r, t)';$$

$$B \cdot (0, 0, 1)' = (u, v, w)'.$$

The above equalities suggest substituting

$$for \quad B \quad the\ matrix \quad \begin{pmatrix} 0 & 0 & 1 \\ q & r & t \\ u & v & w \end{pmatrix}'. \qquad (123)$$

Motivated by (121), (122) and (123), we consider the mapping ψ from $\mathscr{A}$ into the total 3 by 3 matrix algebra $\mathfrak{F}_3$ where for every element S of $\mathscr{A}$ as given by (110), we define

$$\psi(S) = s_1\psi(I) + s_2\psi(A) + s_3\psi(B) \qquad (124)$$

where

$$\psi(I) = \begin{pmatrix} 1 & 0 & 0 \\ 0 & 1 & 0 \\ 0 & 0 & 1 \end{pmatrix}' \quad \psi(A) = \begin{pmatrix} 0 & 1 & 0 \\ a & b & c \\ g & h & k \end{pmatrix}' \quad \psi(B) = \begin{pmatrix} 0 & 0 & 1 \\ q & r & t \\ u & v & w \end{pmatrix}'.$$

$$(125)$$

Using notation introduced in (113), we see that for every element S of $\mathscr{A}$, (121) through (125) yield

$$\psi(S) = \mathbf{N}' \quad if\ and\ only\ if \quad \overline{SX'} = \mathbf{N}'\overline{X}' \quad for\ every\ element\ X\ of\ \mathscr{A} \quad (126)$$

where, naturally, $\psi(S)$ as well as $\mathbf{N}'$ is a 3 by 3 matrix.

From (126) and the fact that the algebra $\mathscr{A}$ (whose multiplication table is given by (105)) has a unity, it follows (as in the case of φ on page 104) that the mapping ψ is an algebra isomorphism from $\mathscr{A}$ onto a subalgebra of $\mathfrak{F}_3$.

DEFINITION 10. *Let $\mathscr{A}$ and $\mathscr{A}'$ be algebras over a field $\mathfrak{F}$ and let θ be a one-to-one mapping from $\mathscr{A}$ onto $\mathscr{A}'$ with $\theta(X) = X'$. Then $\mathscr{A}'$ is* *called reciprocal to $\mathscr{A}$ over $\mathfrak{F}$ if*

$$(X+Y)' = X' + Y', \qquad (XY)' = Y'X', \qquad (sX)' = sX'$$

for every element X and Y of $\mathscr{A}$ and every element s of $\mathfrak{F}$.

Let us again consider the three dimensional algebra $\mathscr{A}$ over $\mathfrak{F}$ where the multiplication table of $\mathscr{A}$ with respect to the basis $\{I, A, B\}$ of $\mathscr{A}$ is given by table (105). Let $\mathscr{S}'$ be the subalgebra of $\mathfrak{F}_3$ generated by the three linearly independent matrices

$$\begin{pmatrix} 1 & 0 & 0 \\ 0 & 1 & 0 \\ 0 & 0 & 1 \end{pmatrix}, \qquad \begin{pmatrix} 0 & 1 & 0 \\ a & b & c \\ g & h & k \end{pmatrix}, \qquad \begin{pmatrix} 0 & 0 & 1 \\ q & r & t \\ u & v & w \end{pmatrix}. \tag{127}$$

Let us consider the mapping θ from $\mathscr{A}$ into $\mathfrak{F}_3$ where for every element S of $\mathscr{A}$, as given by (110), we define

$$\theta(S) = s_1\theta(I) + s_2\theta(A) + s_3\theta(B) \tag{128}$$

where

$$\theta(I) = \begin{pmatrix} 1 & 0 & 0 \\ 0 & 1 & 0 \\ 0 & 0 & 1 \end{pmatrix}, \qquad \theta(A) = \begin{pmatrix} 0 & 1 & 0 \\ a & b & c \\ g & h & k \end{pmatrix}, \qquad \theta(B) = \begin{pmatrix} 0 & 0 & 1 \\ q & r & t \\ u & v & w \end{pmatrix}. \tag{129}$$

As shown on page 107, the mapping ψ is an isomorphism from $\mathscr{A}$ onto the subalgebra $\mathscr{S}$ of $\mathfrak{F}_3$ generated by the three linearly independent matrices mentioned in (121), (122) and (123). However, these matrices are respectively the transposes of the matrices given in (127). Moreover, since $(\mathbf{PQ})' = \mathbf{Q}'\mathbf{P}'$ for every n by n matrices $\mathbf{P}$ and $\mathbf{Q}$, we see at once that θ is a one-to-one mapping from $\mathscr{S}$ onto $\mathscr{S}'$ and θ satisfies the conditions set forth in Definition 10. Therefore, algebra $\mathscr{S}'$ is reciprocal to $\mathscr{S}$ as well as to $\mathscr{A}$. Furthermore, since $(\mathbf{N}')' = \mathbf{N}$ for every matrix $\mathbf{N}$, in view of (126), we see that for every element S of $\mathscr{A}$ we have

$$\theta(S) = \mathbf{N} \quad \textit{if and only if } \overline{SX} = \overline{X}\mathbf{N} \quad \textit{for every element } X \textit{ of } \mathscr{A} \tag{130}$$

REMARK 3. *The reader is advised to observe the simple method by which the matrices given in (127) are constructed from table (105).*

DEFINITION 11. *The representation of reciprocal algebra $\mathscr{A}'$ to an n dimensional algebra $\mathscr{A}$ with unity, by means of n by n matrices as given by (128) and (129) is called the regular representation of $\mathscr{A}'$ with respect to the basis of $\mathscr{A}$ which is under consideration.*

Next, let us observe that if $\mathbf{M}$ is an element of the regular representation of $\mathscr{A}$ and $\mathbf{N}$ an element of the corresponding regular representation of $\mathscr{A}'$ (reciprocal to $\mathscr{A}$) then by (114) and (130) we have

$$(\overline{X}\mathbf{M})\mathbf{N} = (\overline{XS})\mathbf{N} = \overline{S(XS)}$$

and

$$(\overline{X}\mathbf{N})\mathbf{M} = (\overline{SX})\mathbf{M} = \overline{(SX)X}$$

which in view of the associativity of $\mathscr{A}$ and $\mathfrak{F}_3$ imply

$$\overline{X}\mathbf{MN} = \overline{X}\mathbf{NM}$$

for every element X of $\mathscr{A}$. Since in the above equality X can be replaced by $(1, 0, 0)$, $(0, 1, 0)$ and $(0, 0, 1)$ we see that

$$\mathbf{MN} = \mathbf{NM} \tag{131}$$

for every element $\mathbf{M}$ of the regular representation of $\mathscr{A}$ and every element $\mathbf{N}$ of the corresponding regular representation of $\mathscr{A}'$ (reciprocal to $\mathscr{A}$).

Next, let $\mathbf{K}$ be any 3 by 3 matrix over $\mathfrak{F}$ and let $\mathbf{K}$ commute with every element $\mathbf{M}$ of the regular representation of $\mathscr{A}$. Then in view of the commutativity of $\mathbf{M}$ with $\mathbf{K}$ and (114), we have

$$(\overline{X}\mathbf{K})\mathbf{M} = (\overline{X}\mathbf{M})\mathbf{K} = (\overline{XS})\mathbf{K}$$

for every element X and S of $\mathscr{A}$. Since $\mathscr{A}$ has unity I, we may choose in the above $X = I$ and obtain

$$\overline{Y}\mathbf{M} = \overline{S}\mathbf{K}$$

where Y is an element of $\mathscr{A}$ such that $\overline{Y}$ is the first row of $\mathbf{K}$. But from the above equality in view of (114) we derive

$$\overline{YS} = \overline{S}\mathbf{K} \tag{132}$$

for every element S of $\mathscr{A}$. Comparing (132) with (130) we see at once that $\mathbf{K}$ is an element of the regular representation of $\mathscr{A}'$ (reciprocal to $\mathscr{A}$).

Thus, by interchanging in the above the roles of $\mathscr{A}$ and $\mathscr{A}'$, in view of (131) we have proved the following

THEOREM 9. *Let $\mathscr{A}$ be an n dimensional algebra with unity over a field $\mathfrak{F}$. Let $\mathscr{S}$ be the regular representation of $\mathscr{A}$ with respect to a basis of $\mathscr{A}$ and $\mathscr{S}'$ be the corresponding regular representation of reciprocal algebra $\mathscr{A}'$ to $\mathscr{A}$ over $\mathfrak{F}$. Then the centralizer of $\mathscr{S}$ in $\mathfrak{F}_n$ is $\mathscr{S}'$ and vice versa.*

In view of the above theorem, if, for instance, the set of all matrices (with entries in a field $\mathfrak{F}$) of the form

$$
s_1 \begin{pmatrix} 1 & 0 & 0 \\ 0 & 1 & 0 \\ 0 & 0 & 1 \end{pmatrix} + s_2 \begin{pmatrix} 0 & 1 & 0 \\ a & b & c \\ q & r & t \end{pmatrix} + s_3 \begin{pmatrix} 0 & 0 & 1 \\ g & h & k \\ u & v & w \end{pmatrix}
$$

form a subalgebra $\mathscr{S}$ of $\mathfrak{F}_3$ then the centralizer of $\mathscr{S}$ in $\mathfrak{F}_3$ is the set $\mathscr{S}'$ of all matrices of the form

$$
s_4 \begin{pmatrix} 1 & 0 & 0 \\ 0 & 1 & 0 \\ 0 & 0 & 1 \end{pmatrix} + s_5 \begin{pmatrix} 0 & 1 & 0 \\ a & b & c \\ g & h & k \end{pmatrix} + s_6 \begin{pmatrix} 0 & 0 & 1 \\ q & r & t \\ u & v & w \end{pmatrix}
$$

where s_i is an element of $\mathfrak{F}$ for $i = 1, 2, \ldots, 6$. Moreover, the centralizer of $\mathscr{S}'$ in $\mathfrak{F}_3$ is $\mathscr{S}$.

THEOREM 10. *Let $\mathscr{S}$ be an n dimensional algebra with a unity and let the n by n matrix $\mathbf{M}$ be the regular representation of an element S of $\mathscr{A}$ with respect to the basis $\{A_1, A_2, \ldots, A_n\}$ of $\mathscr{A}$. Let $\{B_1, B_2, \ldots, B_n\}$ be a basis of $\mathscr{A}$ such that*

$$
\{B_1, B_2, \ldots, B_n\} = \{A_1, A_2, \ldots, A_n\}\mathbf{P}
$$

where $\mathbf{P}$ is an n by n nonsingular matrix. Then

$$\mathbf{P}^{-1}\mathbf{M}\mathbf{P}$$

is the regular representation of S with respect to the basis $\{B_1, B_2, \ldots, B_n\}$ of $\mathscr{A}$.

Proof. Using notation introduced in (113), in view of (114), we have $\overline{XS} = \overline{X}\mathbf{M}$ for every element X of $\mathscr{A}$. But then

$$\overline{XS}\mathbf{P} = \overline{X}\mathbf{P}(\mathbf{P}^{-1}\mathbf{M}\mathbf{P}). \tag{133}$$

However, $\overline{XS}\mathbf{P}$ and $\overline{X}\mathbf{P}$ are respectively the n-tuples of the coordinates of XS and X with respect to the basis $\{B_1, B_2, \ldots, B_n\}$. Thus, (133), in view of (114), implies that $\mathbf{P}^{-1}\mathbf{M}\mathbf{P}$ is the regular representation of S with respect to the basis $\{B_1, B_2, \ldots, B_n\}$ of $\mathscr{A}$.

REMARK 4. *In view of Theorem 10, we see that in any regular representation of an n dimensional algebra $\mathscr{A}$ with a unity, the unity of $\mathscr{A}$ is represented by the unit n by n matrix.*

Let us give some applications of the regular representation of algebras. First, we introduce the following definition.

DEFINITION 12. *A nonzero idempotent E of an algebra $\mathscr{A}$ is called a principal idempotent of $\mathscr{A}$ if $\mathscr{A}$ has no nonzero idempotent U such that U is orthogonal (i.e., $UE = EU = 0$) to E.*

Clearly, if an algebra $\mathscr{A}$ has a unity element I then I is the only principal idempotent of $\mathscr{A}$.

THEOREM 11. *If E_1 is a nonzero nonprincipal idempotent of an algebra $\mathscr{A}$ then there exists an idempotent E of $\mathscr{A}$ such that E is orthogonal to E_1 and $E_1 + E$ is a principal idempotent of $\mathscr{A}$.*

Proof. In view of Definition 12 there exists a nonzero idempotent E_2 of $\mathscr{A}$ such that $E_1E_2 = E_2E_1 = 0$. Let $\mathbf{E}_1$ and $\mathbf{E}_2$ be n by n matrices respectively corresponding to the elements E_1 and E_2 of a regular representation $\mathscr{R}$ of $\mathscr{A}$. By Corollary 7 on page 100 there exists a nonsingular matrix $\mathbf{P}$ such that

$$\mathbf{P}(\mathbf{E}_1 + \mathbf{E}_2)\mathbf{P}^{-1} = \begin{pmatrix} \mathbf{I}_k & \mathbf{O} \\ \mathbf{O} & \mathbf{O}_{n-k} \end{pmatrix}$$

where $\mathbf{I}_k$ is the k by k unit matrix and $\mathbf{O}_{n-k}$ is the $n-k$ by $n-k$ zero matrix. Let us observe that any (matrix) similarity transformation of $\mathscr{R}$ transforms $\mathscr{R}$ into a subalgebra of the total n by n matrix algebra isomorphic to $\mathscr{A}$. Therefore, if $E_1 + E_2$ is not a principal idempotent of $\mathscr{A}$ then repetitions of the above construction must terminate in finitely many, say, m steps (since rank $\mathbf{E}_1 <$ rank $(\mathbf{E}_1 + \mathbf{E}_2) = k \leqslant n$) implying the existence of a principal idempotent $E_1 + E_2 + \cdots + E_{m+1}$ where, obviously, $E = E_2 + \cdots + E_{m+1}$ is an idempotent orthogonal to E_1.

In view of Theorem 11 we have the obvious

COROLLARY 9. *If an algebra $\mathscr{A}$ has a nonzero idempotent then $\mathscr{A}$ has a principal idempotent.*

DEFINITION 13. *A nonzero idempotent E of an algebra $\mathscr{A}$ is called a primitive idempotent if E is not the sum $E_1 + E_2$ of nonzero orthogonal idempotents E_1 and E_2 of $\mathscr{A}$.*

THEOREM 12. *Every nonzero nonprimitive idempotent E of an algebra $\mathscr{A}$ is the sum of finitely many pairwise orthogonal primitive idempotents of $\mathscr{A}$.*

Proof. Let $E = E_1 + E_2$ where E_1 and E_2 are nonzero orthogonal idempotents of $\mathscr{A}$. Let $\mathbf{E}_1$ and $\mathbf{E}_2$ be n by n matrices respectively corresponding to the elements E_1 and E_2 of a regular representation $\mathscr{R}$ of $\mathscr{A}$. By Corollary 7 on page 100 there exists a nonsingular matrix $\mathbf{P}$ such that

$$\mathbf{P}(\mathbf{E}_1 + \mathbf{E}_2)\mathbf{P}^{-1} = \begin{pmatrix} \mathbf{I}_k & & \mathbf{O} \\ & \mathbf{I}_h & \\ \mathbf{O} & & \mathbf{O}_{n-k-h} \end{pmatrix}$$

where $k =$ rank of $\mathbf{E}_1$ and $h =$ rank $\mathbf{E}_2$ and where $\mathbf{I}_k$ and $\mathbf{I}_h$ are respectively the k by k and h by h unit matrices and $\mathbf{O}_{n-k-h}$ is the $n-h-k$ by $n-h-k$ zero matrix. Again, we observe that any (matrix) similarity transformation of $\mathscr{R}$ transforms $\mathscr{R}$ into a subalgebra of the total n by n matrix algebra isomorphic to $\mathscr{A}$. Thus, if E_1 or E_2 are not primitive idempotents of $\mathscr{A}$ and, say, $E_1 = E_3 + E_4$ and $E_2 = E_5 + E_6$ are pairwise orthogonal nonzero idempotents of $\mathscr{A}$ then repetitions of the above construction must terminate in finitely many steps (since $k + h \leqslant n$) implying the existence of finitely many pairwise orthogonal primitive idempotent E_i of $\mathscr{A}$ whose sum is equal to E.

Combining Theorems 11 and 12, we have:

COROLLARY 10. *Let V be a nonzero nonprincipal idempotent of an algebra $\mathscr{A}$. Then there exists a principal idempotent E of $\mathscr{A}$ and pairwise orthogonal primitive idempotents $E_1, E_2, \ldots, E_m$ of $\mathscr{A}$ such that $E = E_1 + E_2 + \cdots + E_m$ and $V = E_1 + E_2 + \cdots + E_n$ for some $n < m$.*

Let us give some examples of the regular representation of algebras.

Let us consider a two dimensional algebra $\mathscr{C}$ over a field $\mathfrak{F}$ and let $\{I, A\}$ be a basis of $\mathscr{C}$ with the following multiplication table

	I	A
I	$(1, 0)$	$(0, 1)$
A	$(0, 1)$	$(-1, 0)$

Thus, according to Remark 2 on page 106, the regular representation of $\mathscr{C}$ with respect to the basis $\{I, A\}$ is given by the set of all matrices of the form

$$s_1 \begin{pmatrix} 1 & 0 \\ 0 & 1 \end{pmatrix} + s_2 \begin{pmatrix} 0 & 1 \\ -1 & 0 \end{pmatrix} \tag{134}$$

with $s_i \in \mathfrak{F}$.

If $\mathfrak{F}$ is the field of rational or real numbers, it is easily seen that $\mathscr{C}$ is a field. If $\mathfrak{F}$ is the field of real numbers then $\mathscr{C}$ is (isomorphic to) the algebra (or field) of the familiar *complex numbers*.

Next, let us consider a four dimensional algebra $\mathscr{Q}$ over a field $\mathfrak{F}$ and let

$\{I, A, B, C\}$ be a basis of $\mathscr{D}$ with the following multiplication table

	I	A	B	C
I	$(1, 0, 0, 0)$	$(0, 1, 0, \quad 0)$	$(0, \quad 0, 1, 0)$	$(0, 0, \quad 0, 1)$
A	$(0, 1, 0, 0)$	$(-1, 0, 0, \quad 0)$	$(0, \quad 0, 0, 1)$	$(0, 0, -1, 0)$
B	$(0, 0, 1, 0)$	$(0, 0, 0, -1)$	$(-1, \quad 0, 0, 0)$	$(0, 1, \quad 0, 0)$
C	$(0, 0, 0, 1)$	$(0, 0, 1, \quad 0)$	$(0, -1, 0, 0)$	$(-1, 0, \quad 0, 0)$

Then the regular representation of $\mathscr{D}$ with respect to the basis $\{I, A, B, C\}$ is given by the set of all matrices of the form

$$
s_1 \begin{pmatrix} 1 & 0 & 0 & 0 \\ 0 & 1 & 0 & 0 \\ 0 & 0 & 1 & 0 \\ 0 & 0 & 0 & 1 \end{pmatrix} + s_2 \begin{pmatrix} 0 & 1 & 0 & 0 \\ -1 & 0 & 0 & 0 \\ 0 & 0 & 0 & -1 \\ 0 & 0 & 1 & 0 \end{pmatrix} + s_3 \begin{pmatrix} 0 & 0 & 1 & 0 \\ 0 & 0 & 0 & 1 \\ -1 & 0 & 0 & 0 \\ 0 & -1 & 0 & 0 \end{pmatrix}
$$

$$
+ s_4 \begin{pmatrix} 0 & 0 & 0 & 1 \\ 0 & 0 & -1 & 0 \\ 0 & 1 & 0 & 0 \\ -1 & 0 & 0 & 0 \end{pmatrix} \tag{135}
$$

with $s_i \in \mathfrak{F}$.

If $\mathfrak{F}$ is the field of rational or real numbers, it is easily seen that $\mathscr{D}$ is a division ring (*see* page 10). However, if $\mathfrak{F}$ is the field of complex numbers then $\mathscr{D}$ is not a division ring.

If $\mathfrak{F}$ is the field of rationals or reals or complex numbers then $\mathscr{D}$ is (isomorphic to) respectively, the algebra of the familiar *rational, or, real, or, complex quaternion numbers*.

Finally, let us consider the total 2 by 2 matrix algebra $\mathscr{M}_2$ over a field $\mathfrak{F}$ and let $\{A_{11}, A_{21}, A_{12}, A_{22}\}$ be the basis of $\mathscr{M}_2$ (mentioned in Lemma 15 on page 90) with the following multiplication table

	A_{11}	A_{21}	A_{12}	A_{22}
A_{11}	$(1, 0, 0, 0)$	$(0, 0, 0, 0)$	$(0, 0, 1, 0)$	$(0, 0, 0, 0)$
A_{21}	$(0, 1, 0, 0)$	$(0, 0, 0, 0)$	$(0, 0, 0, 1)$	$(0, 0, 0, 0)$
A_{12}	$(0, 0, 0, 0)$	$(1, 0, 0, 0)$	$(0, 0, 0, 0)$	$(0, 0, 1, 0)$
A_{22}	$(0, 0, 0, 0)$	$(0, 1, 0, 0)$	$(0, 0, 0, 0)$	$(0, 0, 0, 1)$

Then the regular representation of $\mathscr{M}_2$ with respect to the basis $\{A_{11},$

$A_{21}, A_{12}, A_{22}\}$ is given by the set of all matrices of the form

$$s_1 \begin{pmatrix} 1 & 0 & 0 & 0 \\ 0 & 1 & 0 & 0 \\ 0 & 0 & 0 & 0 \\ 0 & 0 & 0 & 0 \end{pmatrix} + s_2 \begin{pmatrix} 0 & 0 & 0 & 0 \\ 0 & 0 & 0 & 0 \\ 1 & 0 & 0 & 0 \\ 0 & 1 & 0 & 0 \end{pmatrix} + s_3 \begin{pmatrix} 0 & 0 & 1 & 0 \\ 0 & 0 & 0 & 1 \\ 0 & 0 & 0 & 0 \\ 0 & 0 & 0 & 0 \end{pmatrix} + s_4 \begin{pmatrix} 0 & 0 & 0 & 0 \\ 0 & 0 & 0 & 0 \\ 0 & 0 & 1 & 0 \\ 0 & 0 & 0 & 1 \end{pmatrix}$$

$$(136)$$

with $s_i \in \mathfrak{F}$.

If $\mathfrak{F}$ is the field of complex numbers then it is easy to verify that the algebra of all matrices of the form (135) is isomorphic to the algebra of all matrices of the form (136). Thus, the algebra of complex quaternions is isomorphic to the total 2 by 2 matrix algebra with complex entries.

Exercises

1. Let $\{A, B, C\}$ be a basis of a three dimensional algebra $\mathscr{A}$ over a field $\mathfrak{F}$ such that

$$AA = A, \quad AB = BA = B, \quad AC = BB = CA = C, \quad BC = CB = CC = 0.$$

Consider the regular representation of $\mathscr{A}$ with respect to the basis $\{A, B, C\}$ and determine the matrices (over $\mathfrak{F}$) corresponding to the elements $A^2, B + C$ and $A + B + C$ of $\mathscr{A}$.

2. Let $\{G, H, K\}$ be a basis of a three dimensional algebra $\mathscr{A}$ over the field of rational numbers such that

$$GG = H, \quad GH = HG = K, \quad HH = HK = KH = KK = 0.$$

Using the regular representation of $\mathscr{A}$ with respect to the basis $\{G, H, K\}$ determine the monic polynomial of least degree which is satisfied by the element $2G + H - K$ of $\mathscr{A}$.

3. Let $\{L, M, N\}$ be a basis of a three dimensional algebra $\mathscr{A}$ over the field of real numbers. Using the regular representation of $\mathscr{A}$ with respect to the basis $\{L, M, N\}$, determine the form by a 4 by 4 matrix which represents an arbitrary element of an algebra $\mathscr{S}_1$ such that $\mathscr{S}_1$ has a unity and $\mathscr{A}$ is isomorphic to a sub-algebra of $\mathscr{S}_1$. Also, determine a 4 by 4 matrix which represents an arbitrary element of the centralizer of $\mathscr{S}_1$ in the total 4 by 4 matrix algebra over the field of real numbers.

4. Motivated by (134) on page 112, prove that if r and s are rational or real numbers then every nonzero matrix of the form

$$\begin{pmatrix} r & s \\ -s & r \end{pmatrix}$$

is nonsingular.

5. Motivated by (135) on page 113, prove that if p, q, r and s are rational or real numbers then every nonzero matrix of the form

$$\begin{pmatrix} p & q & r & s \\ -q & p & -s & r \\ -r & s & p & -q \\ -s & -r & p & q \end{pmatrix}$$

is nonsingular.

6. Consider the set $\mathscr{K}$ of all matrices of the form

$$\begin{pmatrix} a+ib & c+id \\ -c+id & a-ib \end{pmatrix}$$

where the entries are complex numbers. Consider also the set $\mathscr{D}$ of all 4 by 4 matrices (with real number entries) of the form mentioned in Problem 5. Prove that there exists a one-to-one mapping φ from $\mathscr{K}$ onto $\mathscr{D}$ such that φ preserves addition and multiplication.

7. Determine a nonzero 4 by 4 matrix $\mathbf{M}$ of the form mentioned in Problem 5 with complex number entries such that $\mathbf{M}$ is singular.

8. Consider the matrices $\mathbf{M}$ and $\mathbf{N}$ with entries in a field $\mathfrak{F}$ where

$$M = \begin{pmatrix} a & b \\ c & d \end{pmatrix} \quad \text{and} \quad N = \begin{pmatrix} m & n \\ p & q \end{pmatrix}.$$

Prove or disprove that if the set $\mathscr{S}$ of all matrices of the form $u\mathbf{M} + v\mathbf{N}$ (with u and v elements of $\mathscr{F}$) is a two dimensional subalgebra with a unity of the total 2 by 2 matrix algebra $\mathfrak{F}_2$ then the centralizer of $\mathscr{S}$ in $\mathfrak{F}_2$ is the set of all matrices of the form

$$u\begin{pmatrix} a & b \\ m & n \end{pmatrix} + v\begin{pmatrix} c & d \\ p & q \end{pmatrix}.$$

3.6. Division Algebras

Let us observe that if A is a nonzero element of an algebra $\mathscr{A}$ then for an element B of $\mathscr{A}$ an equation of the form

$$AX = B \quad \text{or} \quad XA = B \tag{137}$$

may have no solution or may have one or many solutions in $\mathscr{A}$.

For instance, in the total 2 by 2 matrix algebra over the field of real numbers the equation

$$\begin{pmatrix} 0 & 1 \\ 0 & 0 \end{pmatrix} X = \begin{pmatrix} 0 & 0 \\ 0 & 1 \end{pmatrix}$$

has no solution, whereas, the equation

$$\begin{pmatrix} 0 & 1 \\ 0 & 0 \end{pmatrix} X = \begin{pmatrix} 0 & 1 \\ 0 & 0 \end{pmatrix}$$

has infinitely many solutions.

Clearly, if in an algebra $\mathscr{D}$ with a unity (different from 0) every nonzero element is nonsingular (*i.e.*, has a multiplicative inverse which is necessarily unique) then every equation of the form mentioned in (137) has a unique solution. Such an algebra $\mathscr{D}$ is obviously a division ring (*see* page 16) since the set of all nonzero elements of $\mathscr{D}$ is a group under multiplication.

DEFINITION 14. *An algebra $\mathscr{D}$ with more than one element and with a unity is called a division algebra if every nonzero element of $\mathscr{D}$ is nonsingular.*

Examples of division algebras are numerous. For instance, the algebra of complex numbers (*see* page 112) is a two dimensional (commutative) division algebra over the field of real numbers. Moreover, the algebra of real quaternion numbers (*see* page 113) is a four dimensional (noncommutative) division algebra over the field of real numbers. Furthermore, any finite field is a (commutative) division algebra over its prime field.

Let us mention that in view of Definition 14, we exclude an algebra with a single element 0 from being a division algebra. Thus, the unity of a division algebra is necessarily different from 0. Also, it is obvious that the dimension of a division algebra is greater than zero.

LEMMA 19. *Any division algebra is a simple algebra.*

Proof. Let $\mathscr{D}$ be a division algebra. In view of Definition 12, $\mathscr{D}$ is not a one dimensional zero algebra. Moreover, since every nonzero element of $\mathscr{D}$ is nonsingular we see that $\mathscr{D}$ has no nontrivial left or right ideal (*see* page 11) and, consequently, $\mathscr{D}$ has no nontrivial ideal. Hence, $\mathscr{D}$ is a simple algebra (*see* page 86).

Before proving the next theorem we remind the reader of the definition of the *minimum polynomial* and the *minimum equation* of an element A of an algebra $\mathscr{A}$ (*see* page 73). Moreover, we observe that if $\mathscr{A}$ has a unity then the minimum polynomial of A is the minimum polynomial (the first invariant factor) of the matrix $\mathbf{I}x - \mathbf{A}$ where $\mathbf{A}$ is a matrix which corresponds to A in a regular representation of $\mathscr{A}$. Furthermore, we recall that if $\mathscr{A}$ has a unity then A is nonsingular if and only if the constant term of the minimum polynomial of A is nonzero (*see* pages 62 and 78). Finally, we note that a nonzero nilpotent element (*see* pages 9 and 72) of an algebra $\mathscr{A}$ is a divisor of zero (*see* pages 9 and 32) and hence is a singular element of $\mathscr{A}$ if $\mathscr{A}$ has a unity.

THEOREM 13. *Let $\mathscr{D}$ be an algebra with more than one element. Then $\mathscr{D}$ is a division algebra if and only if $\mathscr{D}$ has no divisor of zero.*

Proof. If $\mathscr{D}$ is a division algebra then obviously $\mathscr{D}$ has no divisor of zero since every divisor of zero is a singular element.

Next, let $\mathscr{D}$ have more than one element and have no divisor of zero. Then, clearly, $\mathscr{D}$ has no nonzero nilpotent element and hence, by Corollary 2, page 78, the algebra $\mathscr{D}$ has a nonzero idempotent element E. But then

$$E(EX-X) = (XE-X)E = 0$$

for every element X of $\mathscr{D}$. However, since $\mathscr{D}$ has no divisor of zero and $E \neq 0$, it follows from the above equalities that $EX = XE = X$ which implies that E is the unity of $\mathscr{D}$. Thus $\mathscr{D}$ has a unity. Moreover, the minimum polynomial $f(x)$ of a nonzero element A of $\mathscr{D}$ must have a nonzero constant term since, otherwise, $f(x)$ will be reducible which would imply that A is a divisor of zero. Consequently, every nonzero element of $\mathscr{D}$ is nonsingular. Hence, $\mathscr{D}$ is a division algebra, as desired.

LEMMA 20. *An algebra $\mathscr{D}$ with more than one element over a field $\mathfrak{F}$ is a division algebra if and only if the minimum polynomial of every nonzero element of $\mathscr{D}$ is irreducible in $\mathfrak{F}$.*

Proof. Let $\mathscr{D}$ be a division algebra and $f(x)$ be the minimum polynomial of a nonzero element A of $\mathscr{D}$. If $f(x)$ is reducible and $f(x) = p(x) \cdot h(x)$ then $p(A)h(A) = 0$. However, since $\mathscr{D}$ has no divisor of zero the last equality implies $p(A) = 0$ or $h(A) = 0$ either of which contradicts the fact that $f(x)$ is the minimum polynomial of A. Thus, indeed $f(x)$ is irreducible.

Next, if the minimum polynomial $f(x)$ of a nonzero element A of $\mathscr{D}$ is irreducible in $\mathfrak{F}$ then the constant term of $f(x)$ is nonzero which implies that $\mathscr{D}$ has a unity and A is a nonsingular element of $\mathscr{D}$ (*see* page 78). Thus $\mathscr{D}$ is a division algebra, as desired.

THEOREM 14. *Let $\mathscr{D}$ be an m dimensional division subalgebra of an n dimensional algebra $\mathscr{A}$ with a unity and let the unity of $\mathscr{D}$ be equal to the unity of $\mathscr{A}$. Then m divides n, i.e., $n = mk$ for some natural number k.*

Proof. Let $\mathscr{A}$ be an algebra as described above over a field $\mathfrak{F}$. Since the unity of $\mathscr{D}$ coincides with that of $\mathscr{A}$, we may look at $\mathscr{A}$ as a left module over $\mathscr{D}$ (*see* page 40). But this means that there exist elements $A_1, A_2, \ldots, A_k$ of $\mathscr{A}$ such that every element of $\mathscr{A}$ is a linear combination

$$H_1A_1 + H_2A_2 + \cdots + H_kA_k \tag{138}$$

for some elements H_i of $\mathscr{D}$. Let $\{D_1, D_2, \ldots, D_m\}$ be a basis of $\mathscr{D}$. Thus

$$H_i = s_{i_1}D_1 + s_{i_2}D_2 + \cdots + s_{i_m}D_m \tag{139}$$

for some elements s_{i_j} of $\mathfrak{F}$.

Comparing (138) with (139) we see at once that $\{D_1H_1, D_2H_1, \ldots, D_mH_1, \ldots, D_mH_k\}$ is a basis of $\mathscr{A}$. However, since the dimension of $\mathscr{A}$ is n we must have $n = mk$, as desired.

COROLLARY 11. *Let $\mathscr{D}$ be a division subalgebra of a division algebra $\mathscr{A}$. Then the unity of $\mathscr{D}$ coincides with that of $\mathscr{A}$ and the dimension of $\mathscr{D}$ divides that of $\mathscr{A}$.*

Proof. Let D be a nonzero element of $\mathscr{D}$. Thus, D is a nonzero element of $\mathscr{A}$. Hence, the minimum polynomial of D has a nonzero constant term. Therefore, the unity of $\mathscr{A}$ can be expressed as a linear combination of powers of D (*see* page 78). Consequently, the unity of $\mathscr{A}$ is an element of $\mathscr{D}$ and therefore it is the unity of $\mathscr{D}$. The rest of the proof follows from Theorem 14.

Next, we prove the so-called Frobenius theorem

THEOREM 15. *The one dimensional algebra of real numbers and the two dimensional algebra of complex numbers are the only (finite dimensional) commutative division algebras over the field of real numbers. Moreover, the four dimensional algebra of quaternion numbers is the only (finite dimensional) noncommutative division algebra over the field of real numbers.*

Proof. Let $\mathscr{D}$ be a division algebra over the field of real numbers $\mathfrak{F}$. Let us suppose that $\mathscr{D}$ is neither a one dimensional algebra isomorphic to $\mathfrak{F}$ nor a two dimensional algebra isomorphic to the algebra of complex numbers (*see* page 112). We shall prove that $\mathscr{D}$ is a four dimensional algebra isomorphic to the algebra of real quaternions (*see* page 112).

In view of our supposition the dimension of $\mathscr{D}$ is at least 2. We first show that the dimension of $\mathscr{D}$ is greater than 2. Assume the contrary and let $\{I, D\}$ be a basis of $\mathscr{D}$ where I is the unity of $\mathscr{D}$. By Lemma 20, the minimum polynomial of D is irreducible over $\mathfrak{F}$ and since I and D are linearly independent and $\mathfrak{F}$ is the field of real numbers we see that the minimum polynomial of D must be an irreducible monic polynomial of degree 2, say

$$x^2 + ax + b \qquad \text{with} \qquad 4b - a^2 > 0.$$

Thus,

$$D^2 + aD + bI = 0 \qquad \text{with} \qquad 4b - a^2 > 0. \tag{140}$$

Consider the element A of $\mathcal{Q}$ determined by

$$A = \frac{2}{\sqrt{4b-a^2}}\left(D+\frac{a}{2}I\right). \tag{141}$$

Then, clearly, $\{I, A\}$ is a basis of $\mathcal{Q}$ and in view of (140) and (141) we see that $A^2 = -I$. Thus, $\mathcal{Q}$ is isomorphic to the two dimensional algebra of complex numbers (*see* page 112) contrary to our supposition. Hence, indeed, the dimension of $\mathcal{Q}$ is greater than 2. Consequently, in view of the above considerations there exist three linearly independent elements I, A, G of $\mathcal{Q}$ such that

$$A^2 = G^2 = -I. \tag{142}$$

From the linear independence of I, A and G it follows (as in the case of D) that the minimum polynomial of $A+G$ as well as $A+G$ is an irreducible monic polynomial of degree 2. Thus, there exist elements r, s, u and v of $\widetilde{\mathfrak{F}}$ such that

$$(A+G)^2 = r(A+G) + sI = (AG+GA) - 2I \tag{143}$$

and

$$(A-G)^2 = u(A-G) + vI = -(AG+GA) - 2I \tag{144}$$

where the last equalities in (143) and (144) are obtained by virtue of (142).

Upon adding the equalities in (143) and (144) we obtain

$$(4+s+v)I + (r+u)A + (r-u)G = 0$$

which in view of the linear independence of I, A and G implies

$$r = u = 0 \qquad \text{and} \qquad s+v = -4 \tag{145}$$

But then from (143) it follows that

$$AG + GA = (s+2)I = -(v+2)I. \tag{146}$$

Moreover, from (143) and (144) in view of (145) it follows that

$$s < 0 \qquad \text{and} \qquad v < 0 \tag{147}$$

since the minimum polynomials of A and G are irreducible. Thus, from (146) and (147) we derive that $-4 < s < 0$ and, consequently,

$$-s^2 - 4s > 0.$$

Next, we consider the element B of $\mathcal{Q}$ determined by

$$B = \frac{s+2}{\sqrt{-s^2-4s}}A + \frac{2}{\sqrt{-s^2-4s}}G. \tag{148}$$

Clearly, I, A and B are linearly independent since I, A and G are linearly independent. Also, from (146) and (148) it follows that

$$B^2 = -I \quad \text{and} \quad AB + BA = 0. \tag{149}$$

Now, we consider the element C of $\mathscr{Q}$ determined by

$$C = AB = -BA \tag{150}$$

and we show that I, A, B and C are linearly independent. Indeed, if for p, q and t elements of $\mathfrak{F}$

$$C = pI + qA + tB \tag{151}$$

then in view of (142), (149), (150) and (151) we have

$$AC = A^2B = -B = pA + qA^2 + tAB = t(pI + qA + tB)$$

which implies $p = q = t = 0$ since I, A and B are linearly independent. Thus, I, A, B and C are linearly independent. Moreover, from (142), (149) and (150) we derive

$$\begin{aligned} A^2 = B^2 = C^2 = -I, \quad & AB = -BA = C, \\ BC = -CB = A, \quad & AC = -CA = -B \end{aligned} \tag{152}$$

which shows (since I is the unity of $\mathscr{Q}$) that $\{I, A, B, C\}$ is the basis of a subalgebra $\mathscr{S}$ of $\mathscr{Q}$ such that $\mathscr{S}$ is isomorphic to the four dimensional algebra of the real quaternion numbers (*see* page 112).

Finally, we prove that $\mathscr{S} = \mathscr{Q}$. Let M be an element of $\mathscr{Q}$. Without loss of generality we may assume (as in the case of G) that

$$M^2 = -I$$

Therefore, (as in the case of G) in view of (146) there exist elements k, m and n of $\mathfrak{F}$ such that

$$AM + MA = kI, \quad BM + MB = mI, \quad CM + MC = nI.$$

But then from (150) and the above we obtain

$$\begin{aligned} MC &= (MA)B = kB - A(MB) = kB - mA + (AB)M \\ &= kB - mA + CM = kB - mA + nI - MC \end{aligned}$$

which implies

$$2MC = kB - mA + nI.$$

Multiplying both sides of the above equality on the right by C in view of (152) we obtain

$$2M = -(kA + mB + nC)$$

showing that M is an element of $\mathscr{S}$. Thus, indeed $\mathscr{S} = \mathscr{D}$ and $\mathscr{D}$ is isomorphic to the four dimensional algebra of real quaternion numbers.

Thus, Theorem 15 is proved.

Let us observe that in view of Theorem 15, there exists, for instance, no eight dimensional division algebra over the field of real numbers. In this connection let us mention the following. Let $\mathscr{V}$ be a finite dimensional vector space over the field of real numbers. Let multiplication of vectors in $\mathscr{V}$ be defined in such a way that in the resulting algebraic system $(\mathscr{V}, +, \cdot)$ all the axioms of an algebra are satisfied except for the associativity of the multiplication of vectors which, however, is replaced by the requirement that every system of $(\mathscr{V}, +, \cdot)$ generated by two elements of $\mathscr{V}$ be associative. Such a system is called a finite dimensional *alternative algebra* (*see* page 12). Then it has been shown that the familiar eight dimensional algebra of *Cayley numbers* is the only finite dimensional alternative algebra over the field of real numbers which has no divisor of zero.

Exercises

1. Let $\mathscr{S}$ be an s dimensional subspace of an algebra $\mathscr{A}$ and let $\mathscr{D}$ be an m dimensional division subalgebra of $\mathscr{A}$ whose unity coincides with that of $\mathscr{A}$. Prove that if $\mathscr{D}\mathscr{S} \subset \mathscr{S}$ then m divides s.

2. Let the minimum polynomial of an element A of a division algebra $\mathscr{D}$ be of degree m. Prove that the subalgebra (A) of $\mathscr{D}$ generated by A is an m-dimensional commutative division subalgebra of $\mathscr{D}$.

3. Let $\mathscr{D}$ be a noncommutative division algebra of dimension 4. Prove that the minimum polynomial of a nonzero element of $\mathscr{D}$ is of degree 1 or 2.

4. Let $\mathscr{D}$ be a noncommutative division algebra of dimension 4. Let A and B be linearly independent elements of $\mathscr{D}$ whose minimum polynomials are of degree 2. Prove that $\mathscr{D}$ is generated by A and B, i.e., $(A, B) = \mathscr{D}$ (*see* page 83).

5. Let $\mathscr{A}$ be an algebra with a unity I over the field of real numbers and let $\mathscr{A}$ be generated by I, A and B. Prove that if

$$A^3 - 3A + I = 0, \qquad B^3 = 2I, \qquad AB = B(A^2 - 2I)$$

then $\mathscr{A}$ is a division algebra of dimension 9.

CHAPTER 4

Wedderburn Structure Theorems

4.1. Direct Sum and Tensor Product Decomposition of Algebras

As in the case of rings (*see* Section 1.3), in the case of algebras also there are many ways to form a new algebra from one or more given algebras, and, conversely, to decompose judiciously a given algebra into some component algebras. In the former case, the new algebras may acquire some properties which are not shared by the given algebras. In the latter case, the component algebras may be of simpler structure facilitating the study of the given algebra.

Since an algebra is a ring, the general ideas (developed in Section 1.3) in connection with forming the direct sum or the tensor product of some given rings are applicable to the case of algebras. Similarly, the general ideas (mentioned in Section 1.3) in connection with decomposing a given ring into a direct sum of some of its ideals or decomposing a ring into a tensor product of some of its subrings are also applicable to the case of algebras.

Let us recall again that an algebra, besides being a ring, is also a vector space. Thus, in addition to the above, the general ideas in connection with forming the supplementary sum of vector spaces (*see* page 47) and decomposing a vector space into the supplementary sum of some of its subspaces (*see* Section 2.2) are also applicable to the case of algebras.

Let $\mathscr{A}_1, \mathscr{A}_2, \ldots, \mathscr{A}_n$ be algebras over the same field $\mathfrak{F}$ and let $\mathscr{A}$ be the set of all ordered n-tuples $(A_1 A_2, \ldots, A_n)$ with $A_i \in \mathscr{A}_i$. Then, as in the case of rings (*see* page 15), the *direct sum* or, more appropriately, the *external direct sum*

$$\mathscr{A}_1 \oplus \mathscr{A}_2 \oplus \cdots \oplus \mathscr{A}_n = \bigoplus_{i=1}^{n} \mathscr{A}_i \tag{1}$$

of the algebras $\mathscr{A}_1, \mathscr{A}_2, \ldots, \mathscr{A}_n$ is defined as the set $\mathscr{A}$ where addition and multiplication among the elements of $\mathscr{A}$ are performed coordinate-wise and multiplication of an element (*i.e.*, scalar) s of $\mathfrak{F}$ by an element

$(A_1, A_2, \ldots, A_n)$ of $\mathscr{A}$ is performed according to

$$s(A_1, A_2, \ldots, A_n) = (sA_1, sA_2, \ldots, sA_n).$$

It is easy to verify that (1) with the abovementioned operations is an algebra over $\widetilde{\mathfrak{F}}$ and that

$$\dim \bigoplus_{i=1}^{n} \mathscr{A}_i = \sum_{i=1}^{n} \dim \mathscr{A}_i \qquad (2)$$

Also, it is easy to verify that $\bigoplus_{i=1}^{n} \mathscr{A}_i$ in (1) has a unity if and only if each component (or direct summand) $\mathscr{A}_i$ has a unity.

Let us also observe that if a regular representation of an algebra $\mathscr{A}$ over a field $\widetilde{\mathfrak{F}}$ is given by the set $\{\mathbf{A} \mid \ldots\}$ of n by n matrices $\mathbf{A}$ and a regular representation of an algebra $\mathscr{B}$ over $\widetilde{\mathfrak{F}}$ is given by the set $\{\mathbf{B} \mid \ldots\}$ of m by m matrices $\mathbf{B}$ then a regular representation of $\mathscr{A} \oplus \mathscr{B}$ is given by the set of all $(n+m)$ by $(n+m)$ matrices of the form

$$\begin{pmatrix} \mathbf{A} & \mathbf{O} \\ \mathbf{O} & \mathbf{B} \end{pmatrix}.$$

Motivated by the notion of the external direct sum of algebras (as in the case of rings, *see* page 16) we introduce the following

DEFINITION 1. *An algebra $\mathscr{A}$ is said to be the internal direct sum or, simply, the direct sum of its subalgebras $\mathscr{A}_1, \mathscr{A}_2, \ldots, \mathscr{A}_n$ if*

every element of $\mathscr{A}$ is uniquely represented as a sum of elements each belonging to a distinct $\mathscr{A}_i$ for (3)

$$i = 1, 2, \ldots, n$$

and

for every $A_i \in \mathscr{A}_i$ and $A_j \in \mathscr{A}_j$ for $i, j = 1, 2, \ldots, n$ (4)

$$A_i A_j = 0 \qquad for \qquad i \neq j$$

where 0 is the zero of the algebra.

Here again, if an algebra $\mathscr{A}$ is the direct sum of its subalgebras $\mathscr{A}_1, \mathscr{A}_2, \ldots, \mathscr{A}_n$, we denote this by

$$\mathscr{A} = \mathscr{A}_1 \oplus \mathscr{A}_2 \oplus \cdots \oplus \mathscr{A}_n = \bigoplus_{i=1}^{n} \mathscr{A}_i \qquad (5)$$

and each $\mathscr{A}_i$ is called a *component* or a *direct summand* of $\mathscr{A}$.

Clearly, if (5) holds then $\mathscr{A}$ is the sum (*see* page 16) as well as the supplementary sum (*see* page 47) of its subspaces $\mathscr{A}_1, \mathscr{A}_2, \ldots, \mathscr{A}_n$. Moreover, (2) also holds for (5).

As in the case of rings, it is readily verified that conditions (3) and (4) imply that in (5) every direct summand is an ideal of $\mathscr{A}$. Furthermore, the important Lemmas 1, 2, 3, 4 and 5 (*see* pages 17–19) are valid for the case of algebras.

As an example, we rephrase Lemma 2 on page 18 as follows.

THEOREM 1. *Let $\mathscr{A}$ be an algebra and $\mathscr{I}$ an ideal of $\mathscr{A}$ with a unity. Then $\mathscr{I}$ is a direct summand of $\mathscr{A}$, i.e.,*

$$\mathscr{A} = \mathscr{I} \oplus \mathscr{K}$$

for some ideal $\mathscr{K}$ of $\mathscr{A}$. Moreover, $\mathscr{K}$ is unique.

As in the case of Lemma 2 on page 18, the ideal $\mathscr{K}$ is the set of all elements X of $\mathscr{A}$ such that $XY = YX = 0$ for every element Y of $\mathscr{I}$.

Let us call an algebra $\mathscr{A}$ *reducible* if $\mathscr{A}$ is expressible as a direct sum of two of its proper ideals. Otherwise, we call $\mathscr{A}$ an *irreducible* algebra.

THEOREM 2. *Let $\mathscr{A}$ be a reducible algebra with a unity. Then $\mathscr{A}$ is expressible as a direct sum of irreducible direct summands uniquely except for the order of the direct summands.*

Proof. Since $\mathscr{A}$ is reducible and finite dimensional, $\mathscr{A} = \mathscr{A}_1 \oplus \mathscr{A}_2 \oplus \cdots \oplus \mathscr{A}_n$ where $\mathscr{A}_i$ for $i = 1, 2, \ldots, n$ is an irreducible ideal of $\mathscr{A}$. Let also, $\mathscr{A} = \mathscr{B}_1 \oplus \mathscr{B}_2 \oplus \cdots \oplus \mathscr{B}_m$ where $\mathscr{B}_i$ for $i = 1, 2, \ldots, m$ is an irreducible ideal of $\mathscr{A}$. But then, $\mathscr{B}_i$ is an ideal of $\mathscr{A}$ and by Lemma 4 on page 19

$$\mathscr{B}_i = \mathscr{B}_{i_1} \oplus \mathscr{B}_{i_2} \oplus \cdots \oplus \mathscr{B}_{i_n}$$

where $\mathscr{B}_{i_j}$ is an ideal of $\mathscr{A}_j$. But then, since $\mathscr{B}_i$ is irreducible $\mathscr{B}_i = \mathscr{B}_{i_k}$ for some $k = 1, 2, \ldots, n$. Thus, $\mathscr{B}_i$ is an ideal of $\mathscr{A}_k$ for some k. However, $\mathscr{B}_i$ has a unity (since $\mathscr{A}$ has a unity) and since $\mathscr{A}_k$ is irreducible, by Theorem 1, we must have $\mathscr{B}_i = \mathscr{A}_k$. Similarly, we prove that every $\mathscr{A}_i$ is equal to some $\mathscr{B}_h$. Finally, since $\mathscr{A}_i \neq \mathscr{A}_j$ and $\mathscr{B}_i \neq \mathscr{B}_j$ for $i \neq j$ we see that the theorem is proved.

Let us observe that in view of Definition 1, an algebra $\mathscr{A}$ is a direct sum of its subalgebras $\mathscr{A}_1, \mathscr{A}_2, \ldots, \mathscr{A}_n$ if and only if

$$\mathscr{A} = \mathscr{A}_1 \dotplus \mathscr{A}_2 \dotplus \cdots \dotplus \mathscr{A}_n$$

and the product of every element of a basis of $\mathscr{A}_i$ with every element of a basis of $\mathscr{A}_j$ with $i \neq j$ is equal to zero. In this connection, we have the following.

LEMMA 1. *If an algebra $\mathscr{A}$ has a basis $\{B_1, B_2, \ldots, B_n\}$ whose elements are pairwise orthogonal then*

$$\mathscr{A} = [B_1] \oplus [B_2] \oplus \cdots \oplus [B_n].$$

Proof. Clearly, every element of $\mathscr{A}$ is uniquely expressible as a sum of elements each belonging to the subspace $[B_i]$ of $\mathscr{A}$ generated by B_i. Moreover, since the elements of $\{B_1, B_2, \ldots, B_n\}$ are pairwise orthogonal, *i.e.*, $B_i B_j = 0$ for $i \neq j$ we see that condition (4) of Definition 1 is satisfied. Thus the lemma is proved.

Let an algebra $\mathscr{A}$ be the direct sum of two of its subalgebras $\mathscr{B}$ and $\mathscr{C}$.

Then in view of Theorem 8 on page 102 and Corollary 8 on page 106 there exists an isomorphism of $\mathscr{A}$ into a total matrix algebra where each element of $\mathscr{A}$ is represented by a square matrix of the form

$$\begin{pmatrix} \mathbf{B} & \mathbf{O} \\ \mathbf{O} & \mathbf{O} \end{pmatrix} \quad \text{or} \quad \begin{pmatrix} \mathbf{O} & \mathbf{O} \\ \mathbf{O} & \mathbf{C} \end{pmatrix}$$

or as a sum of matrices of the form above, where $\mathbf{B}$ is some m by m matrix with $\dim \mathscr{B} \leq m \leq 1 + \dim \mathscr{B}$ and $\mathscr{C}$ is some n by n matrix with $\dim \mathscr{C} \leq n \leq 1 + \dim \mathscr{C}$.

Next, let $\mathscr{A}$ be an algebra and E a nonzero idempotent of $\mathscr{A}$. Without loss of generality, let us assume that $\mathscr{A}$ has a regular representation $\mathscr{R}'$ by means of 4 by 4 matrices. Since E is a nonzero idempotent, by Lemma 17 on page 97, there exists a nonsingular 4 by 4 matrix $\mathbf{P}$ such that $\mathscr{R}'$ is transformed (by the similarity $\mathbf{PXP^{-1}}$) into an algebra $\mathscr{R}$ isomorphic to $\mathscr{A}$ and such that the matrix corresponding to E acquires the form, say,

$$\mathbf{E} = \begin{pmatrix} 1 & 0 & 0 & 0 \\ 0 & 1 & 0 & 0 \\ 0 & 0 & 0 & 0 \\ 0 & 0 & 0 & 0 \end{pmatrix}. \tag{6}$$

Now, if

$$\mathbf{A} = \begin{pmatrix} a & b & c & d \\ e & g & h & k \\ p & q & r & s \\ t & u & v & w \end{pmatrix} \tag{7}$$

is an element of $\mathscr{R}$ we see that

$$\mathbf{AE} = \begin{pmatrix} a & b & 0 & 0 \\ e & g & 0 & 0 \\ p & q & 0 & 0 \\ t & u & 0 & 0 \end{pmatrix} \tag{8}$$

is an element of $\mathscr{R}$ and, consequently,

$$\mathbf{A} - \mathbf{AE} = \begin{pmatrix} 0 & 0 & c & d \\ 0 & 0 & h & k \\ 0 & 0 & r & s \\ 0 & 0 & v & w \end{pmatrix} \qquad (9)$$

is an element of $\mathscr{R}$. In view of (7), (8), (9) and notation (55) on page 80, we see that

$$\mathscr{A} = \mathscr{A}E \dotplus \mathscr{L}_E \qquad (10)$$

where, as (8) shows, $\mathscr{A}E$ is the set of all *left invariants* of E (i.e., $XE = X$ for $X \in \mathscr{A}$), and, as (9) shows, $\mathscr{L}_E$ is the set of all *left annihilators* of E (i.e., $XE = 0$ for $X \in \mathscr{A}$).

Obviously, in (10) the set $\mathscr{A}E$ as well as $\mathscr{L}_E$ is a left ideal of $\mathscr{A}$ and therefore, by Lemma 9 on page 82, each is a subalgebra of $\mathscr{A}$.

The supplementary sum decomposition (10) of $\mathscr{A}$ is called the *left-sided Peirce decomposition of $\mathscr{A}$ relative to the idempotent E*.

Again, from (6) and (7), it follows that

$$\mathbf{EA} = \begin{pmatrix} a & b & c & d \\ e & g & h & k \\ 0 & 0 & 0 & 0 \\ 0 & 0 & 0 & 0 \end{pmatrix} \qquad (11)$$

is an element of $\mathscr{R}$ and, consequently,

$$\mathbf{A} - \mathbf{EA} = \begin{pmatrix} 0 & 0 & 0 & 0 \\ 0 & 0 & 0 & 0 \\ p & q & r & s \\ t & u & v & w \end{pmatrix} \qquad (12)$$

is an element of $\mathscr{R}$. Thus, in view of (7), (11) and (12), we have

$$\mathscr{A} = E\mathscr{A} \dotplus \mathscr{R}_E \qquad (13)$$

where, as (11) shows, $E\mathscr{A}$ is the set of all *right invariants* of E (i.e., $EX = X$ for $X \in \mathscr{A}$), and, as (12) shows, $\mathscr{R}_E$ is the set of all *right annihilators* of E (i.e., $EX = 0$ for $X \in \mathscr{A}$).

The supplementary sum decomposition (13) of $\mathscr{A}$ into its two subalgebras $E\mathscr{A}$ and $\mathscr{R}_E$ is called the *right-sided Peirce decomposition of $\mathscr{A}$ relative to the idempotent E*.

127

Finally, let us observe that from (6) and (7) it follows that

$$\mathbf{EAE} = \begin{pmatrix} a & b & 0 & 0 \\ e & g & 0 & 0 \\ 0 & 0 & 0 & 0 \\ 0 & 0 & 0 & 0 \end{pmatrix} \tag{14}$$

is an element of $\mathscr{R}$ and, consequently, in view of (11), (8) and (12) we see that each of the following is an element of $\mathscr{R}$.

$$\mathbf{E(A-AE)} = \begin{pmatrix} 0 & 0 & c & d \\ 0 & 0 & h & k \\ 0 & 0 & 0 & 0 \\ 0 & 0 & 0 & 0 \end{pmatrix} \tag{15}$$

$$\mathbf{(A-EA)E} = \begin{pmatrix} 0 & 0 & 0 & 0 \\ 0 & 0 & 0 & 0 \\ p & q & 0 & 0 \\ t & u & 0 & 0 \end{pmatrix} \tag{16}$$

and

$$\mathbf{A-EA-AE+EAE} = \begin{pmatrix} 0 & 0 & 0 & 0 \\ 0 & 0 & 0 & 0 \\ 0 & 0 & r & s \\ 0 & 0 & v & w \end{pmatrix}. \tag{17}$$

Thus, (7), (14), (15), (16) and (17) show that

$$\mathscr{A} = E\mathscr{A}E \dotplus E\mathscr{L}_E \dotplus \mathscr{R}_E E \dotplus \mathscr{B}_E \tag{18}$$

where $\mathscr{L}_E$ and $\mathscr{R}_E$ are as given by (10) and (13) and where, as (17) shows, $\mathscr{B}_E$ is the set of all *left and right annihilators of E* (i.e., $XE = EX = 0$ for $X \in \mathscr{A}$).

In the supplementary sum (18) of $\mathscr{A}$ each summand is a subalgebra of $\mathscr{A}$. Moreover, as (15) and (16) show $E\mathscr{L}_E$ as well as $\mathscr{R}_E E$ is a zero algebra.

The supplementary sum decomposition (18) *of $\mathscr{A}$ is called the two-sided Peirce decomposition of $\mathscr{A}$ relative to the idempotent E.*

Next, we prove

LEMMA 2. *Let E be a principal idempotent of an algebra $\mathscr{A}$. Then in the two-sided Peirce decomposition of $\mathscr{A}$ relative to E every element of subalgebra $\mathscr{B}_E$ is nilpotent.*

Proof. In view of Corollary 2 on page 78, it is enough to show that algebra $\mathscr{B}_E$ has no nonzero idempotent. Assume on the contrary that

$\mathscr{B}_E$ has a nonzero idempotent. But then as (17) shows $\mathscr{A}$ will have a non-zero idempotent orthogonal to E contradicting the fact that E is a principal idempotent of $\mathscr{A}$.

Thus, the lemma is proved.

On pages 19 and 20 we mentioned the concept of the tensor product of subrings of a ring and the conditions for a ring to be a tensor product of some of its subrings. Motivated by this concept, we introduce first the concept of the external tensor product of two algebras.

Let $\mathscr{A}$ and $\mathscr{B}$ be two algebras over the same field $\mathfrak{F}$ with unity elements I and I' respectively and with no common element.

Let $\{A_1, A_2, \ldots, A_m\}$ be a basis of $\mathscr{A}$ and $\{B_1, B_2, \ldots, B_n\}$ be a basis of $\mathscr{B}$. Then we define the *external tensor product* $\mathscr{A} \otimes \mathscr{B}$ of the algebras $\mathscr{A}$ and $\mathscr{B}$ as an mn dimensional algebra over $\mathfrak{F}$ whose basis elements are the following mn *symbolic products*

$$A_1B_1, \quad A_1B_2, \quad A_1B_3, \ldots, A_mB_n. \tag{19}$$

Multiplication in $\mathscr{A} \otimes \mathscr{B}$ is defined by the following multiplication table for basis (19).

$$(A_uB_v)(A_hB_k) = \left(\sum_i s_iA_i \right)\left(\sum_j t_jB_j \right) = \sum_{i,j} s_it_jA_iB_j \tag{20}$$

where

$$\sum_{i=1}^m s_iA_i = A_uA_h$$

is given by the algebra $\mathscr{A}$ and

$$\sum_{j=1}^m t_jB_j = B_vB_k$$

is given by the algebra $\mathscr{B}$.

Thus, in the definition of the external tensor product $\mathscr{A} \otimes \mathscr{B}$ it is assumed that the following equalities hold among the symbolic products

$$A_iB_j = B_jA_i \qquad \begin{cases} i = 1, 2, \ldots, m \\ j = 1, 2, \ldots, n. \end{cases} \tag{21}$$

REMARK 1. *In view of (19) and (20) we may look upon $\mathscr{A} \otimes \mathscr{B}$ as the "algebra" $\mathscr{A}$ whose "scalar domain" is the algebra $\mathscr{B}$. In particular, if the algebra $\mathscr{B}$ is a field then $\mathscr{A} \otimes \mathscr{B}$ is an m dimensional algebra over the field $\mathscr{B}$. Clearly, $\mathscr{A} \otimes \mathscr{B}$ can be looked upon also as the "algebra" $\mathscr{B}$ whose "scalar domain" is the algebra $\mathscr{A}$.*

Since each $\mathscr{A}$ and $\mathscr{B}$ has a unity element, it follows that $\mathscr{A} \otimes \mathscr{B}$ has a subalgebra isomorphic to $\mathscr{A}$ (namely, the subalgebra whose basis is $\{A_1I', A_2I', \ldots, A_mI'\}$) and a subalgebra isomorphic to $\mathscr{B}$ (namely, the subalgebra whose basis is $\{IB_1, IB_2, \ldots, IB_n\}$). Obviously, II' is the unity of $\mathscr{A} \otimes \mathscr{B}$.

129

There is an easily detectable relationship between the regular representation of $\mathscr{A} \otimes \mathscr{B}$ with respect to the basis (19) and the regular representations of $\mathscr{A}$ and $\mathscr{B}$ with respect to the bases $\{A_1, A_2, \ldots, A_m\}$ and $\{B_1, B_2, \ldots, B_n\}$.

For instance, let $\mathscr{A}$ be a two dimensional algebra with a unity and with basis $\{A_1, A_2\}$ and $\mathscr{B}$ be a three dimensional algebra with a unity and with basis $\{B_1, B_2, B_3\}$ and let

$$\mathbf{A} = \begin{pmatrix} a & b \\ c & d \end{pmatrix} \quad \text{and} \quad \mathbf{B} = \begin{pmatrix} m & n & p \\ q & r & s \\ t & u & v \end{pmatrix} \tag{22}$$

stand respectively for the regular representation of an arbitrary element of $\mathscr{A}$ and that of an arbitrary element of $\mathscr{B}$. Then the regular representation $\mathscr{R}$ of the external tensor product $\mathscr{A} \otimes \mathscr{B}$ with respect to the basis $\{A_1 B_1, A_1 B_2, A_1 B_3, A_2 B_1, A_2 B_2, A_2 B_3\}$ of $\mathscr{A} \otimes \mathscr{B}$ is given by the set of all 6 by 6 matrices of the form

$$\mathbf{K} = \begin{pmatrix} a\begin{pmatrix} m & n & p \\ q & r & s \\ t & u & v \end{pmatrix} & b\begin{pmatrix} m & n & p \\ q & r & s \\ t & u & v \end{pmatrix} \\ c\begin{pmatrix} m & n & p \\ q & r & s \\ t & u & v \end{pmatrix} & d\begin{pmatrix} m & n & p \\ q & r & s \\ t & u & v \end{pmatrix} \end{pmatrix} \tag{23}$$

Moreover, the subset of all matrices of the form

$$\mathbf{C} = \begin{pmatrix} a & 0 & 0 & b & 0 & 0 \\ 0 & a & 0 & 0 & b & 0 \\ 0 & 0 & a & 0 & 0 & b \\ c & 0 & 0 & d & 0 & 0 \\ 0 & c & 0 & 0 & d & 0 \\ 0 & 0 & c & 0 & 0 & d \end{pmatrix} \tag{24}$$

is a subalgebra of $\mathscr{R}$ isomorphic to $\mathscr{A}$ and the set of all matrices of the form

$$\mathbf{M} = \begin{pmatrix} m & n & p & 0 & 0 & 0 \\ q & r & s & 0 & 0 & 0 \\ t & u & v & 0 & 0 & 0 \\ 0 & 0 & 0 & m & n & p \\ 0 & 0 & 0 & q & r & s \\ 0 & 0 & 0 & t & u & v \end{pmatrix} \tag{25}$$

is a subalgebra of $\mathscr{R}$ isomorphic to $\mathscr{B}$.

As expected, it is easily verified that matrices (24) and (25) commute.

REMARK 2. *The reader is advised to observe the method by which matrix* **K** *given in* (23) *is constructed in terms of matrices* **A** *and* **B** *given in* (22).

An *mp* by *nq* matrix **K** which is constructed from an *m* by *n* matrix **A** and a *p* by *q* matrix **B** according to the method by which **K** given in (23) is constructed from matrices **A** and **B** given in (22), as expected, is called the *tensor product* or the *Kronecker product* of matrices **A** and **B** and is denoted by **A** ⊗ **B**. Thus, for matrices given by (22) and (23) we have

$$\mathbf{K} = \mathbf{A} \otimes \mathbf{B} \tag{26}$$

Moreover, for matrix **C** given by (24) and matrix **A** given by (22) we have

$$\mathbf{C} = \mathbf{A} \otimes \mathbf{I}_3 \tag{27}$$

where $\mathbf{I}_3$ is the 3 by 3 unit matrix.

Furthermore, for matrix **M** given by (25) and matrix **B** given by (22) we have

$$\mathbf{M} = \mathbf{I}_2 \otimes \mathbf{B} \tag{28}$$

where $\mathbf{I}_2$ is the 2 by 2 unit matrix.

In view of (26) we see that the regular representation of the external direct product $\mathscr{A} \otimes \mathscr{B}$ (with respect to the basis (19)) of the algebras $\mathscr{A}$ and $\mathscr{B}$ given on page 129 is the set of all matrices **A** ⊗ **B** where **A** and **B** are respectively elements of the regular representations of $\mathscr{A}$ and $\mathscr{B}$ with respect to the bases $\{A_1, A_2, \ldots, A_m\}$ of $\mathscr{A}$ and $\{B_1, B_2, \ldots, B_n\}$ of $\mathscr{B}$.

Let us mention again that if **A** is an *m* by *n* matrix and **B** is a *p* by *q* matrix then **A** ⊗ **B** is an *mp* by *nq* matrix.

With the appropriate assumptions on the dimensions of matrices **A**, **B**, **C** and **D** and matrix **O** (which for our purposes below is a suitable rectangular matrix all of whose entries are 0) each with entries in the same field $\widetilde{\mathfrak{F}}$, we have

$$\mathbf{A} \otimes \mathbf{O} = \mathbf{O} \otimes \mathbf{A} = \mathbf{O} \tag{29}$$

$$\mathbf{I}_m \otimes \mathbf{I}_n = \mathbf{I}_n \otimes \mathbf{I}_m = \mathbf{I}_{mn} \tag{30}$$

where $\mathbf{I}_u$ is the unit *u* by *u* matrix.

$$s\mathbf{A} \otimes t\mathbf{B} = st(\mathbf{A} \otimes \mathbf{B}) \tag{31}$$

where *s* and *t* are scalars (*i.e.*, elements of $\widetilde{\mathfrak{F}}$)

$$(\mathbf{A} \otimes \mathbf{B})(\mathbf{C} \otimes \mathbf{D}) = \mathbf{AC} \otimes \mathbf{BD} \tag{32}$$

$$(\mathbf{A} \otimes \mathbf{B})^{-1} = \mathbf{A}^{-1} \otimes \mathbf{B}^{-1} \tag{33}$$

$$\mathbf{A} \otimes (\mathbf{B}+\mathbf{C}) = (\mathbf{A} \otimes \mathbf{B}) + (\mathbf{A} \otimes \mathbf{C}) \tag{34}$$

$$(\mathbf{A}+\mathbf{B}) \otimes \mathbf{C} = (\mathbf{A} \otimes \mathbf{C}) + (\mathbf{B} \otimes \mathbf{C}) \tag{35}$$

$$\mathbf{A} \otimes \mathbf{B} = (\mathbf{A} \otimes \mathbf{I}_p)(\mathbf{I}_n \otimes \mathbf{B}) = (\mathbf{I}_m \otimes \mathbf{B})(\mathbf{A} \otimes \mathbf{I}_q) \tag{36}$$

where **A** is an m by n and **B** a p by q matrix and $\mathbf{I}_u$ is the u by u unit matrix.

From (36) it follows that

$$(\mathbf{A} \otimes \mathbf{I}_3)(\mathbf{I}_2 \otimes \mathbf{B}) = (\mathbf{I}_2 \otimes \mathbf{B})(\mathbf{A} \otimes \mathbf{I}_3) \tag{37}$$

where **A** and **B** are given by (22).

Clearly, (37), states (as mentioned above) that matrices **C** and **M** given by (27) and (28) commute.

REMARK 3. *In the definition of the external tensor product $\mathscr{A} \otimes \mathscr{B}$ of two algebras $\mathscr{A}$ and $\mathscr{B}$ we assumed that each algebra has a unity. As a consequence of this assumption $\mathscr{A} \otimes \mathscr{B}$ contains subalgebras isomorphic to $\mathscr{A}$ and to $\mathscr{B}$. It is clear, however, that in view of (19) and (20) the external tensor product $\mathscr{A} \otimes \mathscr{B}$ can be defined without requiring that $\mathscr{A}$ or $\mathscr{B}$ have unity elements.*

REMARK 4. *As mentioned on page 130, a regular representation of the external tensor product $\mathscr{A} \otimes \mathscr{B}$ is obtained from those of $\mathscr{A}$ and $\mathscr{B}$ by the method of forming the tensor product of matrices. In this connection, let us mention that if a set $\mathscr{S}$ of square matrices represents (isomorphically but not necessarily as a regular representation) $\mathscr{A}$ and a set $\mathscr{P}$ of square matrices represents (isomorphically but not necessarily as a regular representation) $\mathscr{B}$ then $\mathscr{A} \otimes \mathscr{B}$ is again represented (isomorphically but not necessarily as a regular representation) by the tensor product of the elements of $\mathscr{S}$ and $\mathscr{P}$.*

As an application of the above ideas we prove the following lemma.

LEMMA 3. *The external tensor product $\mathscr{M}_m \otimes \mathscr{M}_n$ of a total m by m matrix algebra $\mathscr{M}_m$ and a total n by n algebra $\mathscr{M}_n$ over the same field $\mathfrak{F}$ is a total mn by mn matrix algebra over $\mathfrak{F}$.*

Proof. In view of Remark 4, we take for a basis of $\mathscr{M}_m$ the matrices $\mathbf{M}_{ij}$ with $i, j = 1, 2, \ldots, m$ where $\mathbf{M}_{ij}$ is an m by m matrix with 1 on its i, j entry and 0 everywhere else. Similarly, for a basis of $\mathscr{M}_n$ we take matrices $\mathbf{N}_{hk}$ with $h, k = 1, 2, \ldots, n$ where $\mathbf{N}_{hk}$ is an n by n matrix with 1 on its h, k entry and 0 everywhere else. By definition, the set of all $\mathbf{M}_{ij} \otimes \mathbf{N}_{hk}$ matrices is a basis for $\mathscr{M}_m \otimes \mathscr{M}_n$. Moreover, in view of (32) we have

$$(\mathbf{M}_{ij} \otimes \mathbf{N}_{hk})(\mathbf{M}_{pq} \otimes \mathbf{N}_{st}) = \mathbf{M}_{ij}\mathbf{M}_{pq} \otimes \mathbf{N}_{hk}\mathbf{N}_{st}.$$

From the above it can be easily verified that the multiplication table for matrices $\mathbf{M}_{ij} \otimes \mathbf{N}_{hk}$ is that of the basis of a total mn by mn matrix algebra mentioned on page 90.

Motivated by the external tensor product of algebras we introduce the following

DEFINITION 2.   *Let $\{B_1, B_2, \ldots, B_m\}$ be a basis of a subalgebra $\mathscr{B}$ of an algebra $\mathscr{A}$ and $\{C_1, C_2, \ldots, C_n\}$ be a basis of a subalgebra $\mathscr{C}$ of $\mathscr{A}$. If for $i = 1, 2, \ldots, m$ and $j = 1, 2, \ldots, n$.*

$$B_i C_j = C_j B_i \tag{38}$$

and if

$$B_1 C_1, \quad B_1 C_2, \ldots, B_m C_n \tag{39}$$

is a basis of $\mathscr{A}$ then $\mathscr{A}$ is said to be the internal tensor product or, simply, the tensor product of its subalgebras $\mathscr{B}$ and $\mathscr{C}$.

Here again, to express the fact that $\mathscr{A}$ is the (internal) tensor product of its subalgebras $\mathscr{B}$ and $\mathscr{C}$ we write

$$\mathscr{A} = \mathscr{B} \otimes \mathscr{C}. \tag{40}$$

We observe that in (40) the same notation is used as in the case of the external tensor product of algebras $\mathscr{B}$ and $\mathscr{C}$. However, the context will make it explicit whether we are concerned with the external tensor product of algebras or the internal tensor product of subalgebras.

In view of (39) we see that if (40) holds then

$$\dim \mathscr{A} = (\dim \mathscr{B})(\dim \mathscr{C}). \tag{41}$$

Moreover, in view of (38) we see that if (40) holds then every element of $\mathscr{B}$ commutes with every element of $\mathscr{C}$.

There is also an easily detectable relationship between the regular representation of $\mathscr{A}$ in (40) and the corresponding regular representation of the subalgebras $\mathscr{B}$ and $\mathscr{C}$ of $\mathscr{A}$.

For instance, let $\mathscr{A}$ be a six dimensional algebra with unity which is the direct product of its subalgebras $\mathscr{B}$ and $\mathscr{C}$ both with unity and respectively of dimensions 2 and 3. Then there is a regular representation of $\mathscr{A}$ in which a certain set of matrices of the form (*see* (24))

$$\begin{pmatrix} b_1 & 0 & 0 & b_2 & 0 & 0 \\ 0 & b_1 & 0 & 0 & b_2 & 0 \\ 0 & 0 & b_1 & 0 & 0 & b_2 \\ b_3 & 0 & 0 & b_4 & 0 & 0 \\ 0 & b_3 & 0 & 0 & b_4 & 0 \\ 0 & 0 & b_3 & 0 & 0 & b_4 \end{pmatrix} \tag{42}$$

is isomorphic to the subalgebra $\mathscr{B}$ of $\mathscr{A}$ and a certain set of matrices of the form (*see* (25)).

$$\begin{pmatrix} c_1 & c_2 & c_3 & 0 & 0 & 0 \\ c_4 & c_5 & c_6 & 0 & 0 & 0 \\ c_7 & c_8 & c_9 & 0 & 0 & 0 \\ 0 & 0 & 0 & c_1 & c_2 & c_3 \\ 0 & 0 & 0 & c_4 & c_5 & c_6 \\ 0 & 0 & 0 & c_7 & c_8 & c_9 \end{pmatrix} \qquad (43)$$

is isomorphic to the subalgebra $\mathscr{C}$ of $\mathscr{A}$.

The connection between the product $\mathscr{BC}$ of subsets $\mathscr{B}$ and $\mathscr{C}$ of an algebra $\mathscr{A}$, as introduced on page 80, and the internal tensor product of subalgebras is given by the following lemma.

LEMMA 4. *Let* $\{B_1, B_2, \ldots, B_m\}$ *be a basis of a subalgebra* $\mathscr{B}$ *of an algebra* $\mathscr{A}$ *and let* $\mathscr{C}$ *be a subalgebra of* $\mathscr{A}$. *Then*

$$\mathscr{A} = \mathscr{B} \otimes \mathscr{C}$$

if and only if

$$\mathscr{A} = \mathscr{BC} = \mathscr{CB} \qquad (44)$$

and for $H_i \in \mathscr{C}$

$$\sum_{i=1}^{m} B_i H_i = 0 \quad implies \quad H_i = 0 \quad for \quad i = 1, 2, \ldots, m \qquad (45)$$

Moreover, if $\mathscr{B}$ *has a unit element* I *and* $\mathscr{C}$ *has a unit element* U *then* $I = U$ *and* I *is the unit element of* $\mathscr{B} \otimes \mathscr{C}$.

Proof. The sufficiency follows from Definition 2. To prove the necessity, let $\{C_1, C_2, \ldots, C_n\}$ be a basis for $\mathscr{C}$. Then from (44) it follows that

$$B_i C_j = C_j B_i, \quad i = 1, 2, \ldots, m, j = 1, 2, \ldots, n. \qquad (46)$$

Moreover, from (45) we see that $B_1 C_1, B_1 C_2, \ldots, B_m C_n$ are linearly independent and, in view of (44), they form a basis for $\mathscr{A}$. Thus, by Definition 2, we have $\mathscr{A} = \mathscr{B} \otimes \mathscr{C}$, as desired.

Next, let I be the unity of $\mathscr{B}$ and U be the unity of $\mathscr{C}$. By (44) we see that $IU = UI$ and thus, IU is the unity of $\mathscr{B} \otimes \mathscr{C}$. On the other hand, $IU - I = IU(IU - I) = 0$, so that $IU = I$. Similarly, $IU - U = IU(IU - U) = 0$, so that $IU = U = I$. Thus, indeed $I = U$ and I is the unity of $\mathscr{B} \otimes \mathscr{C}$.

According to Theorem 1 on page 125, an ideal $\mathscr{I}$, with a unity, of an algebra $\mathscr{A}$ is a direct summand of $\mathscr{A}$. A parallel theorem for a tensor product decomposition of an algebra is given below.

THEOREM 2. *Let* $\mathscr{A}$ *be an algebra with a unity* I. *If* $\mathscr{M}$ *is a total matrix subalgebra of* $\mathscr{A}$ *whose unity is* I *then* $\mathscr{M}$ *is a tensor factor of* $\mathscr{A}$, *i.e.,*

$$\mathscr{A} = \mathscr{M} \otimes \mathscr{C}.$$

Moreover, $\mathscr{C}$ is the centralizer of $\mathscr{M}$ in $\mathscr{A}$.

Proof. Let (A_{ij}) with $i, j = 1, 2, \ldots, m$ be a basis of $\mathscr{M}$ as described in Lemma 15 on page 90. Let A be an element of $\mathscr{A}$. Define

$$P_{hk} = \sum_{u=1}^{m} A_{uh} A A_{ku} \qquad (h, k = 1, 2, \ldots, m). \qquad (46a)$$

Then

$$A_{ij} P_{hk} = A_{ih} A A_{kj} = P_{hk} A_{ij} \qquad (i, j, h, k = 1, 2, \ldots, m). \qquad (47)$$

Now, since $I = \sum_i A_{ii}$, from (46a) we derive

$$\sum_{i,j} A_{ij} P_{ij} = \sum_{i,j,u} A_{ij} A_{ui} A A_{ju} = \sum_{i,u} A_{ii} A A_{uu} = A. \qquad (48)$$

However, from (47) it follows that P_{hk} are elements of the centralizer $\mathscr{C}$ of $\mathscr{M}$ in $\mathscr{A}$ and from (48) it follows that $\mathscr{MC} = \mathscr{CM} = \mathscr{A}$. On the other hand, if $\sum_{i,j} A_{ij} P_{ij} = 0$ then $\sum_u A_{uh} A A_{ku} = 0$ which in view of (46a) implies $P_{hk} = 0$. Consequently, by Lemma 4 we have $\mathscr{A} = \mathscr{M} \otimes \mathscr{C}$, as desired.

It is instructive to give the proof of Theorem 2 in terms of a regular representation of $\mathscr{A}$.

Let $\mathscr{A}$ be an eight dimensional algebra with unity I and let $\mathscr{M}$ be a four-dimensional total 2 by 2 matrix subalgebra of $\mathscr{A}$ whose unity is I. Let (A_{ij}) for $i, j = 1, 2$ be a basis of $\mathscr{M}$ as described in Lemma 15 on page 90. We observe that A_{11} and A_{22} are a pair of orthogonal idempotents and that $A_{11} + A_{22} = I$. Therefore, by Corollary 7 on page 100 and Theorem 10 on page 110, there exists a regular representation $\mathscr{R}$ of $\mathscr{A}$ such that A_{11} and A_{22} are respectively represented by the matrices $\mathbf{A}_{11}$ and $\mathbf{A}_{22}$ (which can easily be proved to be of the same rank) with

$$\mathbf{A}_{11} = \begin{pmatrix} \mathbf{I}_4 & \mathbf{O}_4 \\ \mathbf{O}_4 & \mathbf{O}_4 \end{pmatrix} \qquad \text{and} \qquad \mathbf{A}_{22} = \begin{pmatrix} \mathbf{O}_4 & \mathbf{O}_4 \\ \mathbf{O}_4 & \mathbf{I}_4 \end{pmatrix}$$

where $\mathbf{I}_4$ and $\mathbf{O}_4$ are respectively the 4 by 4 unit and zero matrices.

But then, A_{12} and A_{21} are respectively represented by matrices $\mathbf{A}_{12}$ and $\mathbf{A}_{21}$ with

$$\mathbf{A}_{12} = \begin{pmatrix} \mathbf{O}_4 & \mathbf{I}_4 \\ \mathbf{O}_4 & \mathbf{O}_4 \end{pmatrix} \qquad \text{and} \qquad \mathbf{A}_{21} = \begin{pmatrix} \mathbf{O}_4 & \mathbf{O}_4 \\ \mathbf{I}_4 & \mathbf{O}_4 \end{pmatrix}.$$

Now, let

$$\mathbf{A} = \begin{pmatrix} \mathbf{A}_4 & \mathbf{B}_4 \\ \mathbf{C}_4 & \mathbf{D}_4 \end{pmatrix}$$

be a regular representation of an element A of $\mathscr{A}$, where $\mathbf{A}_4$, $\mathbf{B}_4$, $\mathbf{C}_4$ and $\mathbf{D}_4$ are 4 by 4 matrices.

Clearly,

$$P_{11} = A_{11}AA_{11} + A_{21}AA_{12} = \begin{pmatrix} A_4 & O_4 \\ O_4 & A_4 \end{pmatrix}$$

$$P_{12} = A_{11}AA_{21} + A_{21}AA_{22} = \begin{pmatrix} B_4 & O_4 \\ O_4 & B_4 \end{pmatrix}$$

$$P_{21} = A_{12}AA_{11} + A_{22}AA_{12} = \begin{pmatrix} C_4 & O_4 \\ O_4 & C_4 \end{pmatrix}$$

$$P_{22} = A_{12}AA_{21} + A_{22}AA_{22} = \begin{pmatrix} D_4 & O_4 \\ O_4 & D_4 \end{pmatrix}$$

are elements of $\mathscr{R}$. Moreover, it is obvious that

$$A_{ij}P_{ij} = P_{ij}A_{ij} \qquad (i,j = 1, 2)$$

so that the elements P_{ij} of $\mathscr{A}$ which correspond to P_{ij} of $\mathscr{R}$ are elements of the centralizer $\mathscr{C}$ of $\mathscr{M}$ in $\mathscr{A}$. Furthermore,

$$A_{11}P_{11} + A_{12}P_{12} + A_{21}P_{21} + A_{22}P_{22} = \begin{pmatrix} A_4 & B_4 \\ C_4 & D_4 \end{pmatrix}$$

so that $\mathscr{M}\mathscr{C} = \mathscr{C}\mathscr{M} = \mathscr{A}$.

Finally, we see that

$$\begin{pmatrix} A_4 & B_4 \\ C_4 & D_4 \end{pmatrix} = O_8$$

implies $P_{11} = P_{12} = P_{21} = P_{22} = O_8$.

Consequently, by Lemma 4, we have $\mathscr{A} = \mathscr{M} \otimes \mathscr{C}$, as desired.

Exercises

1. Prove that if $\mathscr{M}$ is a total matrix algebra over a field $\mathfrak{F}$ and $\mathscr{D}$ is a commutative division algebra over $\mathfrak{F}$ then $\mathscr{M} \otimes \mathscr{D}$ is a total matrix algebra over $\mathscr{D}$.

2. Prove that if $\mathscr{M}$ is a total matrix algebra over a field $\mathfrak{F}$ and $\mathscr{D}$ is a division algebra over $\mathfrak{F}$ then $\mathscr{M} \otimes \mathscr{D}$ is a simple algebra over $\mathfrak{F}$.

3. Prove (29) to (37) on page 131.

4. Let **A** and **B** be two square matrices. Prove that $\det(A \otimes B) = (\det A)^m (\det B)^n$ and $\text{tr}(A \otimes B) = (\text{tr } A)(\text{tr } B)$.

5. Prove that an algebra $\mathscr{A}$ with a unity is a direct product of $\mathscr{A}$ with a one dimensional subalgebra of $\mathscr{A}$.

6. Let $\mathscr{C}$ be the two dimensional algebra of complex numbers over the reals. Is $\mathscr{C} \otimes \mathscr{C}$ isomorphic to the algebra of the real quaternions?

4.2.   Nilpotent Algebras and the Radical of an Algebra

As mentioned on page 82, if $\mathscr{A}$ is an algebra then the descending chain of ideals

$$\mathscr{A} \supset \mathscr{A}^2 \supset \mathscr{A}^3 \supset \cdots$$

must terminate yielding a least positive natural number s such that

$$\mathscr{A}^{s-1} = \mathscr{A}^s.$$

If in the above $\mathscr{A}^s \neq \{0\}$ then $\mathscr{A}$ is called a *potent* algebra of index s and if $\mathscr{A}^s = \{0\}$ then $\mathscr{A}$ is called a nilpotent algebra of index s according to the following definition.

DEFINITION 3.   *An algebra $\mathscr{A}$ is called nilpotent of index s if*

$$\mathscr{A}^s = \{0\} \qquad and \qquad \mathscr{A}^{s-1} \neq \{0\}$$

for some positive natural number s.

In the above Definition, the symbol $\mathscr{A}^s$ represents (*see* page 81) the set (which is necessarily an ideal of $\mathscr{A}$) of all the finite sums of the s-termed products of elements of $\mathscr{A}$. We observe also that a nilpotent algebra of index 1 is the zero dimensional algebra $\{0\}$.

Clearly, if an algebra is not nilpotent then it is a potent algebra.

In connection with Definition 3, let us mention that a zero algebra (*see* page 86) of positive dimension is a nilpotent algebra of index 2.

Let us give some examples. Consider the two dimensional algebra $\mathscr{P}$ whose multiplication table with respect to its basis $\{P_1, P_2\}$ is given by

	P_1	P_2
P_1	P_1	P_1
P_2	P_1	P_1

It is easily seen that $\mathscr{P} \neq \mathscr{P}^2 = \mathscr{P}^3$ and therefore is a potent algebra of index 2.

Again, consider the two dimensional algebra $\mathscr{N}$ whose multiplication table with respect to its basis $\{N_1, N_2\}$ is given by

	N_1	N_2
N_1	N_2	0
N_2	0	0

It is easily seen that $\mathscr{N}^2 \neq \{0\}$ and $\mathscr{N}^3 = \{0\}$. Thus, $\mathscr{N}$ is a nilpotent algebra of index 3.

LEMMA 5. *Every subalgebra of a nilpotent algebra is nilpotent.*

Proof. Let $\mathscr{B}$ be a subalgebra of a nilpotent algebra $\mathscr{N}$ of index s. Clearly, $\mathscr{B}^s \subset \mathscr{N}^s = \{0\}$ and therefore $\mathscr{B}$ is a nilpotent algebra, as desired.

THEOREM 3. *A nilpotent algebra $\mathscr{N}$ has a nontrivial ideal if and only if $\mathscr{N}$ is not a zero algebra of dimension $\leqslant 1$.*

Proof. If $\mathscr{N}$ has a nontrivial ideal then $\mathscr{N}$ is of dimension greater than 1. On the other hand, let $\mathscr{N}$ be not a zero algebra of dimension $\leqslant 1$. But then since $\mathscr{N}$ is nilpotent it must be of dimension greater than 1 (*see* Theorem 2, page 86). Let $\{B_1, B_2 \ldots, B_n\}$ be a basis of $\mathscr{N}$. If $\mathscr{N}^2 = \{0\}$ then $(B_2, \ldots, B_n)$ is a nontrivial ideal of $\mathscr{N}$. Otherwise, $\mathscr{N}^2$ is a nontrivial ideal of $\mathscr{N}$.

Let us observe that, in view of Definition 3 on page 86 and Theorem 3 above, we have

COROLLARY 1. *An algebra $\mathscr{A}$ is simple if and only if $\mathscr{A}$ is nonnilpotent and has no nontrivial ideal.*

Thus, we may take Corollary 1 as a definition of a simple algebra.

LEMMA 6. *The sum $\mathscr{L} + \mathscr{K}$ of nilpotent left (right) ideals $\mathscr{L}$ and $\mathscr{K}$ of an algebra $\mathscr{A}$ is a nilpotent left (right) ideal of $\mathscr{A}$.*

Proof. In view of (51) on page 80 and (72) on page 85, we see that $\mathscr{L} + \mathscr{K}$ is a left (right) ideal of $\mathscr{A}$. Now, let h and k be the indices of $\mathscr{L}$ and $\mathscr{K}$ respectively. But then since $\mathscr{L}$ and $\mathscr{K}$ are left (right) ideals $(\mathscr{L} + \mathscr{K})^{h+k-1} = \{0\}$ which implies that $\mathscr{L} + \mathscr{K}$ is a left (right) nilpotent ideal of $\mathscr{A}$.

LEMMA 7. *If $\mathscr{L}$ is a nilpotent left ideal of an algebra $\mathscr{A}$ then $\mathscr{L}\mathscr{A}$ as well as $\mathscr{L} + \mathscr{L}\mathscr{A}$ is a nilpotent ideal of $\mathscr{A}$.*

Proof. Clearly, $\mathscr{L}\mathscr{A}$ is an ideal of $\mathscr{A}$. Moreover, if h is the index of $\mathscr{L}$ then $(\mathscr{L}\mathscr{A})^h \subset \mathscr{L}^h\mathscr{A} = \{0\}$ so that $\mathscr{L}\mathscr{A}$ is a nilpotent ideal of $\mathscr{A}$. But then, by Lemma 6, we see that $\mathscr{L} + \mathscr{L}\mathscr{A}$ is a nilpotent left ideal of $\mathscr{A}$. However, $\mathscr{L} + \mathscr{L}\mathscr{A}$ is an ideal of $\mathscr{A}$ since $(\mathscr{L} + \mathscr{L}\mathscr{A})\mathscr{A} \subset \mathscr{L} + \mathscr{L}\mathscr{A}$. Therefore, $\mathscr{L} + \mathscr{L}\mathscr{A}$ is a nilpotent ideal of $\mathscr{A}$.

LEMMA 8. *Let $\mathscr{A}$ be an algebra. Then $\mathscr{A}$ has a unique nilpotent ideal of maximal dimension.*

Proof. Since $\{0\}$ is a nilpotent ideal of $\mathscr{A}$ there exists a nilpotent ideal $\mathscr{N}$ of $\mathscr{A}$ of maximal dimension. Now, if $\mathscr{R}$ is any other nilpotent ideal of

$\mathscr{A}$ then by Lemma 6 we see that $\mathscr{R} + \mathscr{N}$ is a nilpotent ideal of $\mathscr{A}$ such that $\mathscr{N} \subset \mathscr{R} + \mathscr{N}$. But since $\mathscr{N}$ is of maximal dimension $\mathscr{N} = \mathscr{R} + \mathscr{N}$ which implies the uniqueness of $\mathscr{N}$.

DEFINITION 4.   *The unique nilpotent ideal of maximal dimension of an algebra $\mathscr{A}$ is called the maximal nilpotent ideal of $\mathscr{A}$ or the radical of $\mathscr{A}$.*

In view of the above Definition, we have

COROLLARY 2.   *An algebra $\mathscr{A}$ is nilpotent if and only if $\mathscr{A}$ is its own radical.*

Next, we prove

LEMMA 9.   *Every nilpotent left, right or two-sided ideal of an algebra $\mathscr{A}$ is a subset of the radical of $\mathscr{A}$.*

Proof.   From Lemma 7 and its obvious right-side analogue, it follows that every left (right) nilpotent ideal of $\mathscr{A}$ is a subset of a nilpotent ideal of $\mathscr{A}$. But then by the proof of Lemma 8 we see that every nilpotent ideal of $\mathscr{A}$ is a subset of the radical of $\mathscr{A}$ from which the proof of the lemma follows readily.

DEFINITION 5.   *A nonnilpotent algebra whose radical is $\{0\}$ is called a semi-simple algebra.*

If the radical of an algebra $\mathscr{A}$ is $\{0\}$ then it is customary to call $\mathscr{A}$ an algebra *without radical.*

Thus, *an algebra is semi-simple if and only if it is a nonnilpotent algebra without radical.*

In view of Corollaries 1 and 2, we have

COROLLARY 3.   *A simple algebra is semi-simple.*

Thus, any total matrix algebra and any division algebra is a semi-simple algebra.

LEMMA 10.   *If $\mathscr{A}$ is a nonnilpotent algebra then $\mathscr{A}$ has a nonzero ideal $\mathscr{B}$ such that*

$$\mathscr{B}^2 = \mathscr{B}.$$

Proof.   Since $\mathscr{A}$ is nonnilpotent we see that $\mathscr{A}^s = \mathscr{A}^{s+1} \neq \{0\}$ for some positive natural number s. But then if we take $\mathscr{A}^s = \mathscr{B}$ we see that $\mathscr{B}^2 = \mathscr{B}$, as desired.

Lemma 10 suggests that a nonnilpotent algebra must have a nonzero idempotent element. This is indeed the case as shown below.

THEOREM 4. *If $\mathscr{A}$ is a nonnilpotent algebra then $\mathscr{A}$ has a nonzero idempotent element.*

Proof. If dim $\mathscr{A} = 1$ then by Theorem 2 on page 86 we see that $\mathscr{A}$ has a unit element which obviously is a nonzero idempotent element of $\mathscr{A}$.

Thus, it is enough to prove the theorem for algebras of dimension $\geqslant 2$.

Assume the contrary and let $\mathscr{A}$ be a nonnilpotent algebra of least dimension $n \geqslant 2$ such that $\mathscr{A}$ has no nonzero idempotent element. Let $\{B_1, B_2, \ldots, B_n\}$ be a basis of $\mathscr{A}$. But then

$$\mathscr{A}B_i \subsetneqq \mathscr{A} \qquad (i = 1, 2, \ldots, n) \tag{49}$$

Because, if $\mathscr{A}B_j = \mathscr{A}$ for some B_j then $AB_j = B_j$ for some nonzero element A of $\mathscr{A}$ and consequently $A^2B_j = AB_j$ and $(A^2 - A)B_j = 0$ which by Corollary 3 on page 82 would imply $A^2 = A$, i.e., A would be a nonzero idempotent of $\mathscr{A}$.

Also, from (49) it follows that each left ideal $\mathscr{A}B_i$ for $i = 1, 2, \ldots, n$ must be nilpotent. Because if $\mathscr{A}B_j$ for some B_j is nonnilpotent then $\mathscr{A}B_j$ being of dimension less than n would have a nonzero idempotent element A which would be a nonzero idempotent element of $\mathscr{A}$.

Thus, each $\mathscr{A}B_i$ for $i = 1, 2, \ldots, n$ is a nilpotent left ideal of $\mathscr{A}$, say, of index n_i. But then since $(\mathscr{A}B_i\mathscr{A})^{n_i} \subset (\mathscr{A}B_i)^{n_i} = \{0\}$ we see that each $\mathscr{A}B_i\mathscr{A}$ for $i = 1, 2, \ldots, n$ is a nilpotent ideal of $\mathscr{A}$ and consequently, by Lemma 6,

$$\mathscr{A}B_1\mathscr{A} + \mathscr{A}B_2\mathscr{A} + \cdots + \mathscr{A}B_n\mathscr{A} = \mathscr{A}^3$$

is a nilpotent ideal of $\mathscr{A}$, say, of index k. Thus, $(\mathscr{A}^3)^k = \mathscr{A}^{3k} = \{0\}$ implying that $\mathscr{A}$ is a nilpotent algebra. But this contradicts our assumption. Hence, our assumption is false and the theorem is proved.

Let us observe that if an algebra $\mathscr{A}$ has a nonzero idempotent element then $\mathscr{A}$ cannot be nilpotent. Consequently, in view of Theorem 4, we have

COROLLARY 4. *An algebra is nilpotent if and only if it has no nonzero idempotent element.*

Equivalently,

COROLLARY 5. *An algebra $\mathscr{A}$ is nilpotent if and only if every element of $\mathscr{A}$ is nilpotent.*

Comparing Corollary 4 above with Corollary 2 on page 78, we see that the former is stronger than the latter.

Let us also observe that Corollary 5 states in particular that if every element of an algebra $\mathscr{A}$ is nilpotent then there exists a natural number n such that not only $X^n = 0$ for every element X of $\mathscr{A}$ but any n-termed product $X_1 X_2 \ldots X_n$ of elements $X_1, X_2, \ldots, X_n$ of $\mathscr{A}$ is equal to zero.

If every element of an algebra $\mathscr{A}$ is nilpotent then it is customary to call $\mathscr{A}$ a *nil algebra*. Thus, we may rephrase Corollary 5 as follows

COROLLARY 6. *An algebra is nilpotent if and only if it is a nil algebra.*

As it is readily seen from the proof of Theorem 4, the finite dimensionality of algebras is a key factor in the proofs of Corollaries 4, 5 and 6.

LEMMA 11. *Let A and R be elements of an algebra $\mathscr{A}$. Then AR is a nilpotent element of $\mathscr{A}$ if and only if RA is a nilpotent element of $\mathscr{A}$.*

Proof. If AR is nilpotent then $(AR)^k = 0$ for some positive natural number k. But then $(RA)^{k+1} = R(AR)^k A = 0$ which implies that RA is nilpotent. The converse is proved in a similar way.

DEFINITION 6. *An element R of an algebra $\mathscr{A}$ is called a properly nilpotent element of $\mathscr{A}$ if AR (and consequently RA) is a nilpotent element of $\mathscr{A}$ for every element A of $\mathscr{A}$.*

Clearly, every properly nilpotent element R of an algebra is a nilpotent element since $(RR)^k = 0$ which implies $R^{2k} = 0$. However, the converse is not true as shown by the following example.

$$\begin{pmatrix} 0 & 1 \\ 0 & 0 \end{pmatrix} \begin{pmatrix} 0 & 1 \\ 0 & 0 \end{pmatrix} = \begin{pmatrix} 0 & 0 \\ 0 & 0 \end{pmatrix} \quad \text{and} \quad \begin{pmatrix} 0 & 1 \\ 0 & 0 \end{pmatrix} \begin{pmatrix} 0 & 0 \\ 1 & 0 \end{pmatrix} = \begin{pmatrix} 1 & 0 \\ 0 & 0 \end{pmatrix}$$

where the right hand product is an idempotent (and hence nonnilpotent) matrix.

Next, we characterize the radical of an algebra in terms of its properly nilpotent element.

THEOREM 5. *The radical $\mathscr{N}$ of an algebra $\mathscr{A}$ is the set of all properly nilpotent elements of $\mathscr{A}$.*

Proof. Since $\mathscr{N}$ is a nilpotent ideal of $\mathscr{A}$ we see that if R is an element of $\mathscr{N}$ then AR is an element of $\mathscr{N}$ for every element A of $\mathscr{A}$ and AR is nilpotent. Hence every element of $\mathscr{N}$ is properly nilpotent. Consequently, $\mathscr{N} \subset \mathscr{P}$ where $\mathscr{P}$ is the set of all properly nilpotent elements of $\mathscr{A}$.

Conversely, let P be an element of $\mathscr{P}$. Then, by Corollary 5, the left ideal $\mathscr{A}P$ is nilpotent since every element of $\mathscr{A}P$ is nilpotent. Consequently, by Lemma 9, we see that $\mathscr{A}P \subset \mathscr{N}$. Similarly, if Q is an element of $\mathscr{P}$ then $\mathscr{A}Q \subset \mathscr{N}$. However, $\mathscr{N}$ is an ideal and therefore for every scalar p and q we have $\mathscr{A}(pP+qQ) \subset \mathscr{N}$. But since $\mathscr{N} \subset \mathscr{P}$ we see that $pP+qQ$ is a properly nilpotent element of $\mathscr{A}$. Thus, $\mathscr{P}$ is a subalgebra of $\mathscr{A}$ since the product of two properly nilpotent elements is a properly nilpotent element. But then $\mathscr{A}\mathscr{P} \subset \mathscr{N} \subset \mathscr{P}$ which implies that $\mathscr{A}\mathscr{P}$ is a left ideal of $\mathscr{A}$ and that $\mathscr{P}^2 \subset \mathscr{N}$. Hence, $\mathscr{P}^2$ and consequently $\mathscr{P}$ is a nilpotent left ideal of $\mathscr{A}$ and as such $\mathscr{P} \subset \mathscr{N}$. But then since $\mathscr{N} \subset \mathscr{P}$ we conclude that $\mathscr{P} = \mathscr{N}$, as desired.

Let us recall that in the Peirce decomposition (*see* (18) on page 128) of an algebra $\mathscr{A}$ relative to a principal idempotent E, the subalgebra $\mathscr{B}_E$ is nil (*see* Lemma 2 on page 128) and therefore by Corollary 6 is nilpotent. Moreover, $E\mathscr{L}_E$ as well as $\mathscr{R}_E E$ is a zero algebra. In this connection we prove the following theorem.

THEOREM 6. *Let E be a principal idempotent of an algebra $\mathscr{A}$ and*

$$\mathscr{A} = E\mathscr{A}E \dotplus E\mathscr{L}_E \dotplus \mathscr{R}_E E \dotplus \mathscr{B}_E$$

be the Peirce decomposition of $\mathscr{A}$ relative to E. Then $E\mathscr{L}_E \dotplus \mathscr{R}_E E \dotplus \mathscr{B}_E$ is a subset of the radical of $\mathscr{A}$.

Proof. As (9) and (12) on page 127 show $\mathscr{L}_E$ is a left and $\mathscr{R}_E$ is a right ideal of $\mathscr{A}$. Thus $\mathscr{L}_E \mathscr{R}_E$ is an ideal of $\mathscr{A}$. However, in view of (17) on page 128 we have $\mathscr{R}_E \mathscr{L}_E \subset \mathscr{B}_E$. As mentioned above $\mathscr{B}_E$ is a nilpotent algebra, say, of index k. Thus $(\mathscr{R}_E \mathscr{L}_E)^k = \{0\}$ and therefore $(\mathscr{L}_E \mathscr{R}_E)^{k+1} = \mathscr{L}_E (\mathscr{R}_E \mathscr{L}_E)^k \mathscr{R}_E = \{0\}$. Therefore the ideal $\mathscr{L}_E \mathscr{R}_E$ is nilpotent and is a subset of the radical $\mathscr{N}$ of $\mathscr{A}$. However, as (9) and (12) on page 127 show, $\mathscr{L}_E \mathscr{A} \subset \mathscr{L}_E \mathscr{R}_E$ and $\mathscr{A}\mathscr{R}_E \subset \mathscr{L}_E \mathscr{R}_E$. Thus, $\mathscr{L}_E \mathscr{A} \subset \mathscr{N}$ and $\mathscr{A}\mathscr{R}_E \subset \mathscr{N}$. Consequently, $\mathscr{L}_E^2 \subset \mathscr{N}$ and $\mathscr{R}_E^2 \subset \mathscr{N}$. Thus, $\mathscr{L}_E$ is a nilpotent left ideal and $\mathscr{R}_E$ is a nilpotent right ideal of $\mathscr{A}$ and by Lemma 9 we see that $(\mathscr{L}_E + \mathscr{R}_E) \subset \mathscr{N}$. But clearly, $(E\mathscr{L}_E \dotplus \mathscr{R}_E E \dotplus \mathscr{B}_E) \subset (\mathscr{L}_E + \mathscr{R}_E)$ and hence $E\mathscr{L}_E \dotplus \mathscr{R}_E E \dotplus \mathscr{B}_E \subset \mathscr{N}$, as desired.

THEOREM 7. *Let E be a nonzero idempotent of an algebra $\mathscr{A}$ and let*

$$\mathscr{A} = E\mathscr{A}E \dotplus E\mathscr{L}_E \dotplus \mathscr{R}_E E \dotplus \mathscr{B}_E$$

be the Peirce decomposition of $\mathscr{A}$ relative to E. If $\mathscr{N}$ is the radical of $\mathscr{A}$ then $(E\mathscr{A}E) \cap \mathscr{N} = E\mathscr{N}E$ is the radical of $E\mathscr{A}E$ and $\mathscr{B}_E \cap \mathscr{N}$ is the radical of $\mathscr{B}_E$.

Proof. Since $\mathscr{N}$ is an ideal of $\mathscr{A}$ we see that $(E\mathscr{A}E) \cap \mathscr{N} = E\mathscr{N}E$. In view of Theorem 5, every element of $E\mathcal{N}E$ is a properly nilpotent element of subalgebra $E\mathcal{A}E$. Thus, to prove the first part of the Theorem (in view of Theorem 5) it is enough to show that every properly nilpotent element of subalgebra $E\mathcal{A}E$ is a properly nilpotent element of $\mathcal{A}$.

Now, let A be an element of $\mathcal{A}$. Then a properly nilpotent element of subalgebra $E\mathcal{A}E$ is an element EPE such that

$$(EAEEPE)^k = (EAEPE)^k = 0 \qquad (50)$$

for some positive natural number k. But then since $E^2 = E$, in view of (50), we have

$$(AEPE)^{k+1} = AEP(EAEPE)^k = 0$$

implying that EPE is a properly nilpotent element of $\mathcal{A}$.

Similarly, in view of Theorem 5, to prove the second part of the theorem it is enough to show that every properly nilpotent element of subalgebra $\mathcal{B}_E$ is a properly nilpotent element of $\mathcal{A}$.

Now, let A be an element of $\mathcal{A}$. Then, in view of (17) on page 128, a properly nilpotent element of subalgebra $\mathcal{B}_E$ is an element $P - EP - AP + EPE$ such that

$$((A - EA - AE + EAE)(P - EP - PE + EPE))^k = 0 \qquad (51)$$

for some positive natural number k. But then since $E^2 = E$, in view of (51), it can be readily verified that

$$(A(P - EP - PE + EPE))^{k+1} = 0$$

implying that $P - EP - PE + EPE$ is a properly nilpotent element of $\mathcal{A}$.

The decomposition of the radical of an algebra $\mathcal{A}$ in a direct sum decomposition of $\mathcal{A}$ is given by the following lemma.

LEMMA 12. *Let $\mathcal{A}$ be an algebra with a unit element and let*

$$\mathcal{A} = \mathcal{A}_1 \oplus \mathcal{A}_2 \oplus \cdots \oplus \mathcal{A}_n.$$

Then a subalgebra $\mathcal{N}$ of $\mathcal{A}$ is the radical of $\mathcal{A}$ if and only if

$$\mathcal{N} = \mathcal{N}_1 \oplus \mathcal{N}_2 \oplus \cdots \oplus \mathcal{N}_n$$

where $\mathcal{N}_i$ is the radical of $\mathcal{A}_i$ and $\mathcal{N}_i = \mathcal{N} \cap \mathcal{A}_i$.

Proof. Let $\mathcal{N}$ be the radical of $\mathcal{A}$. Then since $\mathcal{N}$ is nilpotent, by Lemma 4 on page 19 we see that $\mathcal{N}_i = \mathcal{N} \cap \mathcal{A}_i$ is a nilpotent ideal of $\mathcal{A}_i$. But then, by Lemma 9 on page 139, we have $\mathcal{N}_i \subset \mathcal{N}_i'$ where $\mathcal{N}_i'$ is the radical of $\mathcal{A}_i$. However, since $\mathcal{N}_i'$ is a nilpotent ideal of $\mathcal{A}$, again by Lemma 9 on page 139 we have $\mathcal{N}_i' \subset \mathcal{N}$ and by Lemma 4 on page 139 we have

$\mathcal{N}_i' \subset \mathcal{N} \cap \mathcal{A}_i = \mathcal{N}_i$. Thus, $\mathcal{N}_i' = \mathcal{N}_i$. The converse is proved by reversing the steps in the above proof.

Exercises

1. Let $\mathcal{A} = \mathcal{B} \oplus \mathcal{C}$ where $\mathcal{B}$ and $\mathcal{C}$ are simple subalgebras of the algebra $\mathcal{A}$. Prove that $\mathcal{A}$ is semi-simple and not simple.

2. Let E be a nonzero idempotent element of an algebra $\mathcal{A}$. Prove that E is a primitive idempotent of $\mathcal{A}$ if and only if E is the only nonzero idempotent of $E\mathcal{A}E$.

3. Let E be the only nonzero idempotent element of an algebra $\mathcal{A}$. Let A be a nonzero element of $\mathcal{A}$ such that $AX = E$ for no element X of $\mathcal{A}$. Prove that A is a nilpotent element of $\mathcal{A}$.

4. Consider the three dimensional algebra $\mathcal{A}$ over a field $\mathfrak{F}$ and let the multiplication table of $\mathcal{A}$ with respect to the basis $\{A, B, C\}$ of $\mathcal{A}$ be given by:

$$AA = A, \quad AB = BA = B, \quad AC = BB = CA = C, \quad BC = CB = CC = 0.$$

With appropriate assumptions on $\mathfrak{F}$ discuss the potency, nilpotency, semi-simplicity of $\mathcal{A}$.

5. Same as in Problem 4 where the multiplication table of $\mathcal{A}$ is given by:

$$AA = B, \quad AC = -CA = B, \quad AB = BA = BB = BC = CB = CC = 0.$$

6. Same as in Problem 4 where $\mathcal{A}$ is a four dimensional algebra with basis $\{A, B, C, D\}$ and where $AA = A, AB = BA = CD = B, DA = D$ and otherwise the product of every two basis elements is zero.

7. Same as in Problem 4 where $\mathcal{A}$ is a four dimensional algebra with basis $\{A, B, C, D\}$ and where $AA = B, AB = BA = C, AC = BB = CA = D$ and otherwise the product of every two basis elements is zero.

4.3. Structure of Semi-Simple Algebras

As mentioned on page 139 (*see* Definition 5) a semi-simple algebra is a nonnilpotent algebra without radical, *i.e.,* whose radical is $\{0\}$. On the other hand, as mentioned on page 139 (*see* Corollary 1) a simple algebra is a nonnilpotent algebra without nontrivial ideals. Thus, a simple algebra is a special case of a semi-simple algebra (*see* Corollary 3, page 139).

In this section, we prove some theorems concerning semi-simple algebras, and, in particular, we prove the main theorem which states that a semi-simple algebra is either a simple algebra or a direct sum of simple algebras.

THEOREM 8. *Let $\mathcal{N}$ be a nilpotent ideal of a nonnilpotent algebra $\mathcal{A}$. Then the quotient algebra $\mathcal{A}/\mathcal{N}$ is semi-simple if and only if $\mathcal{N}$ is the radical of $\mathcal{A}$.*

Proof. Let $\mathcal{A}/\mathcal{N}$ be semi-simple. If $\mathcal{N}$ is not the radical of $\mathcal{A}$ then by Theorem 5, we see that $\mathcal{N}$ is a proper subset of the radical $\mathcal{R}$ of $\mathcal{A}$. But then by Lemma 13 on page 88, we see that $\mathcal{R}/\mathcal{N}$ is a nontrivial ideal of $\mathcal{A}/\mathcal{N}$. Clearly, $\mathcal{R}/\mathcal{N}$ is nilpotent and this contradicts the fact that $\mathcal{A}/\mathcal{N}$ is semi-simple. Thus, indeed $\mathcal{N}$ is the radical of $\mathcal{A}$.

Next, let $\mathcal{N}$ be the radical of $\mathcal{A}$. If $\mathcal{A}/\mathcal{N}$ is not semi-simple then since $\mathcal{A}/\mathcal{N}$ is nonnilpotent it must have a nontrivial radical $\mathcal{R}/\mathcal{N}$. But then, by Lemma 13 on page 88, we see that $\mathcal{R}$ is an ideal of $\mathcal{A}$ such that $\mathcal{N} \subseteq \mathcal{R}$. Clearly, $\mathcal{R}$ is nilpotent and this, in view of Theorem 5, contradicts the fact that $\mathcal{N}$ is the radical of $\mathcal{A}$. Thus, indeed $\mathcal{A}/\mathcal{N}$ is semi-simple.

THEOREM 9. *Let E be a nonzero idempotent of a semi-simple algebra $\mathcal{A}$. Then $E\mathcal{A}E$ is a semi-simple algebra.*

Proof. Since the radical of $\mathcal{A}$ is $\{0\}$ it follows from Theorem 7 that the radical of $E\mathcal{A}E$ is also $\{0\}$. On the other hand, clearly $E\mathcal{A}E$ is a non-nilpotent algebra and thus it is a semi-simple algebra.

THEOREM 10. *Every semi-simple algebra has a unit element.*

Proof. Let $\mathcal{A}$ be a semi-simple algebra. Since $\mathcal{A}$ is nonnilpotent, by Theorem 4 it has a nonzero idempotent element. But then by Corollary 9 on page 111, the algebra $\mathcal{A}$ has a principal idempotent E. However, since the radical of $\mathcal{A}$ is $\{0\}$, by Theorem 6 we see that $\mathcal{A} = E\mathcal{A}E$. Clearly, E is the unit element of $E\mathcal{A}E$. Consequently, E is the unit element of $\mathcal{A}$.

Since a simple algebra is semi-simple, we have

COROLLARY 7. *Every simple algebra has a unit element.*

Let us recall (*see* page 125) that an algebra $\mathcal{A}$ is called *irreducible* if $\mathcal{A}$ is not a direct sum of two of its proper ideals. In this connection we have the following theorem.

THEOREM 11. *A semi-simple algebra is irreducible if and only if it is simple.*

Proof. Let $\mathcal{A}$ be a simple algebra then by Corollary 1 we see that $\mathcal{A}$ has no nontrivial ideal. Thus, $\mathcal{A}$ is irreducible.

Next, let $\mathcal{A}$ be semi-simple and irreducible. We show that $\mathcal{A}$ is simple. Clearly, it is enough to show that $\mathcal{A}$ has no nontrivial ideal. Assume the contrary, that $\mathcal{A}$ has a nontrivial ideal $\mathcal{B}$. Since $\mathcal{A}$ is semi-simple, $\mathcal{B}$ is nonnilpotent. Let $\mathcal{N}$ be the radical of $\mathcal{B}$. Thus, $\mathcal{B}\mathcal{N}\mathcal{B} \subset \mathcal{N}$ and $\mathcal{B}\mathcal{N}\mathcal{B}$ is nilpotent. However, $\mathcal{A}\mathcal{B} = \mathcal{B}\mathcal{A} = \mathcal{B}$ since $\mathcal{A}$ has a unity by Theorem 10. Therefore, $\mathcal{A}(\mathcal{B}\mathcal{N}\mathcal{B})\mathcal{A} = \mathcal{B}\mathcal{N}\mathcal{B}$. Consequently, $\mathcal{B}\mathcal{N}\mathcal{B}$ is a nilpotent ideal of $\mathcal{A}$ and since $\mathcal{A}$ is semi-simple $\mathcal{B}\mathcal{N}\mathcal{B} = \{0\}$. Clearly, $\mathcal{A}\mathcal{N}\mathcal{A}$ is an ideal of $\mathcal{A}$. Moreover, since $\mathcal{A}\mathcal{B}\mathcal{A} \subset \mathcal{B}$ we see that $\mathcal{A}\mathcal{N}\mathcal{A} \subset \mathcal{B}$. On the other hand,

$$(\mathcal{A}\mathcal{N}\mathcal{A})^3 = (\mathcal{A}\mathcal{N}\mathcal{A})\mathcal{N}(\mathcal{A}\mathcal{N}\mathcal{A}) \subset \mathcal{B}\mathcal{N}\mathcal{B} = \{0\}.$$

Thus, $\mathcal{A}\mathcal{N}\mathcal{A}$ is a nilpotent ideal of $\mathcal{A}$ and since $\mathcal{A}$ is semi-simple $\mathcal{A}\mathcal{N}\mathcal{A} = \{0\}$. But since $\mathcal{A}$ has a unity $\mathcal{N} \subset \mathcal{A}\mathcal{N}\mathcal{A}$. Hence, $\mathcal{N} = \{0\}$. From this it follows that $\mathcal{B}$ is semi-simple and has a unit element. But then by Theorem 1 on page 125, we see that $\mathcal{B}$ is a direct summand of $\mathcal{A}$ contradicting the fact that $\mathcal{A}$ is irreducible. Consequently, our assumption is false and indeed $\mathcal{A}$ is simple, as desired.

From the above proof, we have

COROLLARY 8. *Every nonzero ideal of a semi-simple algebra is semi-simple.*

Next, we prove the main theorem concerning semi-simple algebras.

THEOREM 12. *An algebra is semi-simple if and only if it is simple or is expressible as a direct sum of simple subalgebras uniquely except for the order of the direct summands.*

Proof. Let $\mathcal{A} = \mathcal{S}_1 \oplus \mathcal{S}_2 \oplus \cdots \oplus \mathcal{S}_n$ where each $\mathcal{S}_i$ is a simple subalgebra of $\mathcal{A}$. By Corollary 7, each $\mathcal{S}_i$ has a unit element E_i. Thus $\sum E_i$ is the unit element of $\mathcal{A}$ and therefore $\mathcal{A}$ is not nilpotent. Since each $\mathcal{S}_i$ is simple the radical of each $\mathcal{S}_i$ is $\{0\}$. Hence, by Lemma 12 on page 143, we see that the radical of $\mathcal{A}$ is $\{0\}$. Consequently, $\mathcal{A}$ is a semi-simple algebra.

Conversely, let $\mathcal{A}$ be a semi-simple algebra. If $\mathcal{A}$ is irreducible then by Theorem 11, it is simple. If $\mathcal{A}$ is reducible then, since $\mathcal{A}$ has a unit element, we see by Theorem 2 on page 125, that $\mathcal{A}$ is expressible as a direct sum of irreducible ideals $\mathcal{S}_i$ uniquely except for the order $\mathcal{S}_i$'s. But then by Corollary 8, each irreducible $\mathcal{S}_i$ is semi-simple and therefore by Theorem 11, each $\mathcal{S}_i$ is simple.

Exercises

1. Let $\mathcal{A}$ be a two dimensional algebra over a field $\mathfrak{F}$ such that every element of $\mathcal{A}$ is idempotent. Prove that the characteristic of $\mathfrak{F}$ is 2 and that $\mathcal{A}$ is a commutative semi-simple algebra. Decompose $\mathcal{A}$ into a direct sum of simple subalgebras of $\mathcal{A}$. Is each direct summand isomorphic to $\mathfrak{F}$? Generalize the results.

2. Let $\mathscr{A}$ be a two dimensional algebra over a field $\mathfrak{F}$ whose characteristic is a prime number p. Prove that if $X^p = X$ for every element X of $\mathscr{A}$ then $\mathscr{A}$ is a commutative semi-simple algebra. Decompose $\mathscr{A}$ into a direct sum of simple subalgebras of $\mathscr{A}$. Is each direct summand isomorphic to $\mathfrak{F}$? Generalize the results.

3. Let $\mathscr{A}$ be a two dimensional algebra over a field $\mathfrak{F}$ of characteristic 3 such that $X^2 + X + 2I = 0$ for every nonzero element X of $\mathscr{A}$ where I is the unity of $\mathscr{A}$. Is $\mathscr{A}$ a semi-simple algebra? If so, decompose $\mathscr{A}$ into a direct sum of simple subalgebras of $\mathscr{A}$. Is each direct summand isomorphic to $\mathfrak{F}$?

4. Prove that every nonzero ideal of a semi-simple algebra has a unit element.

5. Prove that the center of a semi-simple algebra is semi-simple.

6. Let $\{E_1, E_2, \ldots, E_n\}$ be a basis of an algebra $\mathscr{A}$ such that $E_i E_j = 0$ if $i \neq j$ and $E_i E_i = E_i$. Prove that $\mathscr{A}$ is a commutative semi-simple algebra.

4.4. Structure of Simple Algebras

As mentioned on page 138 (*see* Corollary 1) a nonnilpotent algebra with no nontrivial ideal is called a simple algebra. Thus, in particular, a semi-simple algebra with no nontrivial ideal is a simple algebra.

In this section, we prove some theorems concerning simple algebras, and, in particular, we prove the main theorem which states that a simple algebra is a tensor product of a total matrix algebra and a division algebra.

THEOREM 13. *Let E be a nonzero idempotent element of a simple algebra $\mathscr{A}$. Then $E\mathscr{A}E$ is simple.*

Proof. By Corollary 3 and Theorem 9 we see that $E\mathscr{A}E$ is semi-simple. To prove the theorem it remains to show that every nontrivial ideal of $E\mathscr{A}E$ is equal to $E\mathscr{A}E$. Let $\mathscr{B}$ be a nontrivial ideal of $E\mathscr{A}E$. Thus, in particular, $(E\mathscr{A}E)\mathscr{B}(E\mathscr{A}E) \subset \mathscr{B}$. Since E is the unit element of $E\mathscr{A}E$ and $\mathscr{B} \subset E\mathscr{A}E$ we have $E\mathscr{B}E = \mathscr{B}$ and therefore

$$(E\mathscr{A}E)\mathscr{B}(E\mathscr{A}E) = E\mathscr{A}\mathscr{B}\mathscr{A}E \subset \mathscr{B}. \tag{52}$$

Since $\mathscr{A}\mathscr{B}\mathscr{A}$ is a nontrivial ideal of $\mathscr{A}$ and $\mathscr{A}$ is simple $\mathscr{A}\mathscr{B}\mathscr{A} = \mathscr{A}$. But this, in view of (52), implies $E\mathscr{A}E \subset \mathscr{B}$. However, $\mathscr{B} = E\mathscr{B}E \subset E\mathscr{A}E$. Hence $\mathscr{B} = E\mathscr{A}E$, as desired.

Based on the notion of a primitive idempotent element of an algebra (*see* Definition 13 on page 111) we prove the following theorem.

THEOREM 14. *Let $\mathscr{A}$ be a simple algebra and E be an idempotent of $\mathscr{A}$. Then $E\mathscr{A}E$ is a division algebra if and only if E is a primitive idempotent element of $\mathscr{A}$.*

Proof. Let $E\mathscr{A}E$ be a division algebra. Clearly, E is the unit element of $E\mathscr{A}E$. Now, if H is a nonzero idempotent of $E\mathscr{A}E$ then since $E\mathscr{A}E$ is a division algebra, we see that $HH = H$ implies $H = E$. Thus, E is the only nonzero idempotent of $E\mathscr{A}E$ and therefore E is a primitive idempotent of $\mathscr{A}$ by Theorem 12 on page 111.

Conversely, let E be a primitive idempotent of $\mathscr{A}$ and let A be a nonzero element of $E\mathscr{A}E$. Then $AE\mathscr{A}E$ is a nonzero right ideal of $E\mathscr{A}E$. However, $E\mathscr{A}E$ is simple (*see* Theorem 13) and therefore its radical is $\{0\}$. Thus, $AE\mathscr{A}E$ is nonnilpotent (*see* Lemma 9) and by Theorem 4 we see that $AE\mathscr{A}E$ has a nonzero idempotent element U. But clearly E is the unit element of $E\mathscr{A}E$ and consequently $U = E$ because otherwise $E = (E - U) + U$ would contradict the fact that E is a primitive idempotent (*see* Definition 13 on page 111). Hence, the unit E of $E\mathscr{A}E$ is an element of $AE\mathscr{A}E$ and therefore $E = AB$ for an element B of $E\mathscr{A}E$. Thus, every nonzero element A of $E\mathscr{A}E$ has a multiplicative inverse and $E\mathscr{A}E$ is a division algebra, as desired.

Motivated by Theorem 2 on page 134, we prove the main theorem concerning simple algebras.

THEOREM 15. *An algebra $\mathscr{A}$ is simple if and only if $\mathscr{A}$ is a tensor product $\mathscr{M} \otimes \mathscr{D}$ of a total matrix algebra $\mathscr{M}$ and a division algebra $\mathscr{D}$.*

Moreover, if $\mathscr{A} = \mathscr{M} \otimes \mathscr{D} = \mathscr{M}' \otimes \mathscr{D}'$, where $\mathscr{M}'$ is a total matrix algebra and $\mathscr{D}'$ a division algebra, then there exists a nonsingular element A of $\mathscr{A}$ such that $\mathscr{M}' = A\mathscr{M}A^{-1}$ and $\mathscr{D}' = A\mathscr{D}A^{-1}$.

Proof. First, we show that if $\mathscr{A} = \mathscr{M} \otimes \mathscr{D}$ then $\mathscr{A}$ is a simple algebra. Let I be the unit element of $\mathscr{M}$ and U that of $\mathscr{D}$. By Lemma 4 on page 134 we see that $I = U$ is the unit element of $\mathscr{M} \otimes \mathscr{D}$. Thus, $\mathscr{A}$ is nonnilpotent and it remains to prove that every nonzero ideal $\mathscr{B}$ of $\mathscr{M} \otimes \mathscr{D}$ is equal to $\mathscr{A}$. To this end, it is enough to show that I is an element of $\mathscr{B}$. As in Lemma 15 on page 90, let $\{A_{11}, \ldots, A_{ij}, \ldots, A_{nn}\}$ be a basis of $\mathscr{M}$. Then (*see* Definition 2 on page 133) if B is a nonzero element of $\mathscr{B}$ we have

$$B = \sum_{i,j} D_{ij}A_{ij} \qquad (i, j = 1, 2, \ldots, n)$$

where each D_{ij} is an element of $\mathscr{D}$. Since $B \neq 0$ we must have $D_{pq} \neq 0$ for some p and q.

Since $\mathscr{B}$ is an ideal we see that

$$\sum_{k} A_{kp}BA_{qk} = \sum_{k,i,j} A_{kp}D_{ij}A_{ij}A_{qk} \tag{53}$$

is an element of $\mathscr{B}$. However, (*see* Definition 2 on page 133) each D_{ij}

commutes with each A_{ij} and therefore (53) reduces to

$$\sum_{k,i,j} A_{kp}A_{ij}A_{qk}D_{ij}. \tag{54}$$

But since $A_{mn}A_{nv} = A_{mv}$ and $A_{mn}A_{uv} = 0$ if $n \ne u$ we see that

$$\sum_k A_{kk}D_{pq} = \left(\sum A_{kk} \right) D_{pq} = ID_{pq} = D_{pq}.$$

Therefore, D_{pq} is an element of $\mathscr{B}$. Since $\mathscr{D}$ is a division algebra and $D_{pq} \ne 0$ there exists an element V of $\mathscr{A}$ such that $D_{pq}V = I$. Consequently, $D_{pq}V = I$ is an element of the ideal $\mathscr{B}$ and $\mathscr{A} = \mathscr{B}$. Thus, indeed $\mathscr{A}$ is a simple algebra.

Next, we prove that if $\mathscr{A}$ is a simple algebra then $\mathscr{A} = \mathscr{M} \otimes \mathscr{D}$. By Corollary 7 we see that $\mathscr{A}$ has a unit element E and by Theorem 12 on page 111

$$E = E_1 + E_2 + \cdots + E_m$$

where $E_1, E_2, \ldots, E_m$ are pairwise orthogonal primitive idempotent elements of $\mathscr{A}$. Thus, each E_i is a primitive idempotent and for $i = 1, 2, \ldots, m$

$$E_iE_i = E_i \quad \text{and} \quad E_iE_j = 0 \quad \text{if } i \ne j. \tag{55}$$

Define

$$\mathscr{A}_{ij} = E_i\mathscr{A}E_j \quad (i,j = 1, 2, \ldots, m) \tag{56}$$

We claim that

$$\mathscr{A}_{ij}\mathscr{A}_{jk} = \mathscr{A}_{ik} \quad \text{and} \quad \mathscr{A}_{ij}\mathscr{A}_{hk} = \{0\} \quad \text{if } j \ne h. \tag{57}$$

Because $\mathscr{A}_{ij}\mathscr{A}_{jk} = E_i\mathscr{A}E_jE_j\mathscr{A}E_k = E_i\mathscr{A}E_j\mathscr{A}E_k$ by (55). However, E_j is an element of the ideal $\mathscr{A}E_j\mathscr{A}$ of $\mathscr{A}$ and since $\mathscr{A}$ is a simple algebra and $\mathscr{A}E_j\mathscr{A}$ is not zero, we have $\mathscr{A}E_j\mathscr{A} = \mathscr{A}$. Hence, $\mathscr{A}_{ij}\mathscr{A}_{jk} = E_i\mathscr{A}E_k$ and by (56) we see that $\mathscr{A}_{ij}\mathscr{A}_{jk} = \mathscr{A}_{ik}$. On the other hand, $\mathscr{A}_{ij}\mathscr{A}_{hk} = E_i\mathscr{A}E_jE_h\mathscr{A}E_k$ which by (55) implies that $\mathscr{A}_{ij}\mathscr{A}_{hk} = \{0\}$ for $j \ne h$. Thus, (57) is established.

Let us observe that $\mathscr{A}_{ij}$ is a subalgebra of $\mathscr{A}$ and if A_{ij} is an element of $\mathscr{A}_{ij}$ in view of (55) we have

$$E_iA_{ij} = A_{ij}E_j = A_{ij} \quad \text{and} \quad E_hA_{ij} = A_{ij}E_k = 0 \quad \text{if } h \ne i \text{ and } j \ne k. \tag{58}$$

In view of (55) and (57), it is clear that there exists an element A_{1j} of $\mathscr{A}_{1j}$ and an element E_{j1} of $\mathscr{A}_{j1}$ such that $A_{1j}E_{j1} = A_j \ne 0$ is an element of $\mathscr{A}_{11}$.

However, in view of (56) and Theorem 14, we see that $\mathscr{A}_{11}$ is a division algebra. Therefore, since A_j and E_j are nonzero elements of $\mathscr{A}_{11}$ there

149

exists a nonzero element C_j of $\mathcal{A}_{11}$ such that $C_j A_j = E_1$. But then, in view of (58), we see that $C_j A_{1j} = E_{1j}$ is an element of $\mathcal{A}_{1j}$. Thus, in view of (55), we see that there exist elements E_{1j} of $\mathcal{A}_{1j}$ and elements E_{j1} of $\mathcal{A}_{j1}$ such that

$$E_{1j} E_{j1} = E_{11} = E_1 \neq 0 \qquad (j = 1, 2, \ldots, m) \tag{59}$$

Let us define

$$E_{ij} = E_{i1} E_{1j} \qquad (i, j = 1, 2, \ldots, m) \tag{60}$$

From (58), (59) and (60) it follows that

$$E_{ij} E_{jk} = E_{ik} \quad \text{and} \quad E_{ij} E_{hk} = 0 \qquad \text{if } j \neq h \qquad (i, j, k, h = 1, 2, \ldots, m). \tag{61}$$

Since as (59) shows, $E_1 \neq 0$, we see by (61) and Lemma 15 on page 90, that the m^2 elements E_{ij} of $\mathcal{A}$ form a basis of a total m by m matrix sub-algebra $\mathcal{M}$ of $\mathcal{A}$. However, (61) implies that $(E_{ii})^2 = E_{ii} \neq 0$ and therefore E_{ii} is a nonzero idempotent of $E_i \mathcal{A} E_i$. Consequently, $E_{ii} = E_i$ since $E_i \mathcal{A} E_i$ is a division algebra. But then

$$E_1 + \cdots + E_m = E_{11} + \cdots + E_{mm} = E$$

shows that E is the unit element of both $\mathcal{A}$ and $\mathcal{M}$. Consequently, by Theorem 2 on page 134 we see that $\mathcal{M}$ is a tensor factor of $\mathcal{A}$ and we have $\mathcal{A} = \mathcal{M} \otimes \mathcal{D}$ where $\mathcal{D}$ is the centralizer of $\mathcal{M}$ in $\mathcal{A}$.

Since E_{ij} form a basis of $\mathcal{M}$ we see that

$$\mathcal{A} = E_{11} \mathcal{D} \dotplus E_{12} \mathcal{D} \dotplus \cdots \dotplus E_{mm} \mathcal{D}$$

and $E_{ij} \mathcal{D} = \mathcal{D} E_{ij}$. Thus, in view of (55) we have

$$E_{11} \mathcal{A} E_{11} = E_{11} \mathcal{D}. \tag{62}$$

But then since E_{11} is a nonzero idempotent and $E_{11} D = D E_{11}$ for every element D of $\mathcal{D}$, in view of (45) on page 134 and Corollary 3 on page 82, we see that the mapping φ given by

$$\varphi(D) = E_{11} D \tag{63}$$

is an isomorphism from $\mathcal{D}$ onto $E_{11} \mathcal{D}$. However, $E_{11} \mathcal{A} E_{11}$ is a division algebra (*see* Theorem 14) and therefore (62) and (63) imply that $\mathcal{D}$ is a division algebra, as desired.

Finally, let $\mathcal{A}$ be simple and

$$\mathcal{A} = \mathcal{M} \otimes \mathcal{D} = \mathcal{M}' \otimes \mathcal{D}' \tag{64}$$

as mentioned in the statement of the theorem. We show that there exists a nonsingular element A of $\mathcal{A}$ such that $\mathcal{M}' = A \mathcal{M} A^{-1}$ and $\mathcal{D}' = A \mathcal{D} A^{-1}$.

Without loss of generality we assume that $\mathcal{M}$ is a total 2 by 2 matrix algebra and that $\mathscr{M}'$ is a total m by m matrix algebra with $m \geqslant 2$. Let (E_{ij}) with $i, j = 1, 2$ be the usual basis of $\mathscr{M}$ and (G_{ij}) with $i, j = 1, \ldots, m$ be that of $\mathscr{M}'$. From (64) it follows that

$$G_{11} = D_{11}E_{11} + \cdots + D_{22}E_{22} \tag{65}$$

where each D_{ij} is an element of $\mathscr{D}$ and at least one of them, say, D_{12} is nonzero. Let

$$H = D_{12}^{-1}E_{11}G_{11} \qquad \text{and} \qquad K = G_{11}E_{21}. \tag{66}$$

But then from (65) it follows that

$$HK = E_{11}. \tag{67}$$

However, $HKHK = H(KH)K = E_{11}$ and therefore KH is nonzero. Moreover, by (66) and (67) we see that

$$KHKH = KE_{11}H = G_{11}E_{21}E_{11}H = G_{11}E_{21}H = KH \tag{68}$$

and therefore KH is a nonzero idempotent element. Since, as (66) shows, $KH = G_{11}(E_{21}D_{12}^{-1}E_{11})G_{11}$ we see that KH is an element of the division algebra $G_{11}\mathscr{A}G_{11}$ and since KH is nonzero and idempotent, we have

$$KH = G_{11} \tag{69}$$

because G_{11} is the unit (and therefore the only nonzero idempotent) of $G_{11}\mathscr{A}G_{11}$.

Next, consider

$$A = G_{11}KE_{11} + G_{21}KE_{12} \qquad \text{and} \qquad B = E_{11}HG_{11} + E_{21}HG_{12}. \tag{70}$$

Clearly,

$$BA = E_{11}HG_{11}KE_{11} + E_{21}HG_{11}KE_{12}.$$

However, from (66) we see that $HG_{11} = H$ and thus, in view of (67), we have

$$BA = E_{11} + E_{22} = I \tag{71}$$

where I is the unity of $\mathscr{A}$.

Also, from (70) we have

$$AB = G_{11}KE_{11}HG_{11} + G_{21}KE_{11}HG_{12}.$$

Again, from (66) we have $KE_{11} = G_{11}E_{21}$ and from (68) and (69) we have $G_{11}E_{21}H = KH = G_{11}$ and therefore

$$AB = G_{11} + G_{22}.$$

Comparing the above with (71) we see that

$$I = G_{11} + G_{22}.$$

Therefore, $m = 2$ and consequently $\mathscr{M}'$ is isomorphic to $\mathscr{M}$. Moreover, from (71) we see that $A^{-1} = B$ and therefore, in view of (70), (66), (68) and (69), we have

$$AE_{ij}A^{-1} = AE_{ij}B = G_{ij} \qquad \text{for} \qquad i, j = 1, 2$$

from which it follows that $\mathscr{M}' = A\mathscr{M}A^{-1}$, as desired. Furthermore, since $\mathscr{D}$ is the centralizer of $\mathscr{M}$ in $\mathscr{A}$ and $\mathscr{D}'$ is the centralizer of $\mathscr{M}'$ in $\mathscr{A}$ and since $\mathscr{M}' = A\mathscr{M}A^{-1}$ we see that $\mathscr{D}' = A\mathscr{D}A^{-1}$.

Thus, the theorem is proved.

In accordance with Theorem 15, a total matrix algebra $\mathscr{M}$ over a field $\mathfrak{F}$ is equal to $\mathscr{M} \otimes \mathscr{D}$ where $\mathscr{D}$ is a one dimensional subalgebra of $\mathscr{M}$ isomorphic to $\mathfrak{F}$. Similarly, a division algebra $\mathscr{P}$ over a field $\mathfrak{F}$ is equal to $\mathscr{K} \otimes \mathscr{P}$ where $\mathscr{K}$ is a total one by one matrix subalgebra of $\mathscr{P}$ isomorphic to $\mathfrak{F}$.

COROLLARY 9. *A commutative simple algebra is a commutative division algebra.*

Proof. Let $\mathscr{A}$ be a commutative simple algebra. By Theorem 15 we have $\mathscr{A} = \mathscr{M} \otimes \mathscr{D}$ where $\mathscr{M}$ is a total matrix algebra and $\mathscr{D}$ a division algebra. Since $\mathscr{A}$ is commutative $\mathscr{M}$ is a total one by one matrix algebra and therefore $\mathscr{A} = \mathscr{D}$ which implies that $\mathscr{A}$ is a commutative division algebra.

Clearly, from Corollary 9 and Theorem 12 we have

COROLLARY 10. *A commutative semi-simple algebra is a direct sum of commutative division algebras.*

Recalling the definitions of centralizer, center and a central algebra (*see* pages 83 and 84) we prove the following lamma.

LEMMA 13. *Let $\mathscr{B}$ and $\mathscr{C}$ be subalgebras of an algebra $\mathscr{A}$ over a field $\mathfrak{F}$ such that $\mathscr{B}$ and $\mathscr{C}$ have unit elements and $\mathscr{B}$ is central over $\mathfrak{F}$ and $\mathscr{A} = \mathscr{B} \otimes \mathscr{C}$. Then $\mathscr{C}$ is the centralizer of $\mathscr{B}$ in $\mathscr{A}$ and the center of $\mathscr{A}$ is equal to the center of $\mathscr{C}$.*

Proof. By Lemma 4 on page 134 algebra $\mathscr{A} = \mathscr{B} \otimes \mathscr{C}$ has a unity I which is equal to the unities of $\mathscr{B}$ and $\mathscr{C}$. Moreover, since every element of $\mathscr{C}$ commutes with every element of $\mathscr{B}$ we have $\mathscr{C} \subset \mathrm{czr}\,(\mathscr{B})$. Furthermore, it is clear that $\mathrm{cnt}\,(\mathscr{C}) \subset \mathrm{cnt}\,(\mathscr{A})$. Thus, it remains to show that czr $(\mathscr{B}) \subset \mathscr{C}$ and that cnt $(\mathscr{A}) \subset$ cnt $(\mathscr{C})$. Let $\{C_1, C_2, \ldots, C_m\}$ be a basis of $\mathscr{C}$ and let A be an element of czr $(\mathscr{B})$. However, A as an element of $\mathscr{A} = \mathscr{B} \otimes \mathscr{C}$ is uniquely expressible as $B_1 C_1 + B_2 C_2 + \cdots + B_m C_m$ where each B_i is an element of $\mathscr{B}$. Also $AB = BA$ and hence $C_i B = BC_i$ for every element B of $\mathscr{B}$. But then by Lemma 4 on page 134 we see that $AB - BA = \Sigma \ (B_i B - B B_i) C_i = 0$ if and only if $B_i B = B B_i$ for every element B of $\mathscr{B}$. Therefore, each B_i is an element of cnt $(\mathscr{B})$ and since $\mathscr{B}$ is a central algebra each B_i is equal to $I b_i$ for some element b_i of $\mathfrak{F}$. Hence, A is an element of $\mathscr{C}$ and czr $(\mathscr{B}) \subset \mathscr{C}$. Thus, czr $(\mathscr{B}) = \mathscr{C}$. On the other hand, if H is an element of cnt $(\mathscr{A})$ then H is an element of czr $(\mathscr{B})$ and consequently H is an element of $\mathscr{C}$. But then clearly, H is an element of cnt $(\mathscr{C})$ and therefore cnt $(\mathscr{A}) \subset$ cnt $(\mathscr{C})$. Thus, cnt $(\mathscr{A}) =$ cnt $(\mathscr{C})$, as desired.

Clearly, a total matrix algebra is a central simple algebra (*see* Lemma 8, page 31 and Corollary 1, page 32). Moreover the center of a division algebra $\mathscr{D}$ is a field and hence $\mathscr{D}$ is a central division algebra over its center. Thus, in view of Lemma 13 we have the following corollary.

COROLLARY 11. *Let $\mathscr{A} = \mathscr{M} \otimes \mathscr{D}$ be a simple algebra where $\mathscr{M}$ is a total matrix algebra and $\mathscr{D}$ a division algebra. Then $\mathscr{A}$ is central simple over the center of $\mathscr{D}$ and as such $\mathscr{A}$ is the tensor product of $\mathscr{M}$ and the central division algebra $\mathscr{D}$ over the center of $\mathscr{D}$.*

The above corollary shows an instance of how an algebra $\mathscr{A}$ over a field $\mathfrak{F}$ is effected by an extension $\mathfrak{S}$ of the field $\mathfrak{F}$ (in fact, in general dim $(\mathscr{A}$ over $\mathfrak{S}) \leq$ dim $(\mathscr{A}$ over $\mathfrak{F}))$. In this connection we introduce the following definition.

DEFINITION 7. *Let $\mathfrak{S}$ be a field which is an extension of a field $\mathfrak{F}$ and let $\mathscr{A}$ be an algebra over $\mathfrak{F}$. Then by $\mathscr{A}$ over $\mathfrak{S}$ we mean an algebra over $\mathfrak{S}$ whose multiplication table is that of $\mathscr{A}$ (clearly, a basis of $\mathscr{A}$ over $\mathfrak{F}$ is a set of generators of $\mathscr{A}$ over $\mathfrak{S}$).*

Let us give an example. As mentioned on page 113, the real quaternion algebra $\mathscr{D}$ is a four dimensional division algebra over the field $\mathfrak{F}$ of real numbers. The multiplication table of $\mathscr{D}$ with respect to the basis $\{I, A, B, C\}$ of $\mathscr{D}$ is given by:

$$A^2 = B^2 = C^2 = -I, \qquad AB = -BA = C,$$
$$BC = -CB = A, \qquad CA = -AC = B.$$

It was also mentioned on page 114 that if the real field $\mathfrak{F}$ is extended to the complex field $\mathfrak{S}$ then $\mathscr{D}$ (naturally, with the above multiplication table) is a total 2 by 2 matrix algebra over $\mathfrak{S}$. Clearly, $\mathscr{D}$ over $\mathfrak{F}$ is a division algebra whereas $\mathscr{D}$ over $\mathfrak{S}$ is not a division algebra.

In connection with the above example let us observe that $\mathcal{D}$ over $\mathfrak{F}$ is a central simple algebra and that $\mathfrak{S}$ is a scalar extension of finite degree (namely, 2) over $\mathfrak{F}$. Thus, as a consequence of the scalar extension $\mathfrak{S}$ of $\mathfrak{F}$ of finite degree over $\mathfrak{F}$ the central simple algebra $\mathcal{D}$ over $\mathfrak{F}$ became a total matrix algebra over $\mathfrak{S}$. This fact is a particular instance of the theorem below whose proof follows from the problems given in the subsequent exercises (*see* Exercise 6 below).

THEOREM 17. *Let $\mathcal{A}$ be an algebra over a field $\mathfrak{F}$. Then $\mathcal{A}$ is central simple over $\mathfrak{F}$ if and only if there exists a scalar extension $\mathfrak{S}$ of $\mathfrak{F}$ of finite degree over $\mathfrak{F}$ such that $\mathcal{A}$ is a total matrix algebra over $\mathfrak{S}$.*

In view of Corollary 11 and the above theorem, we have

THEOREM 18. *Let $\mathcal{A}$ be a simple algebra over a field $\mathfrak{F}$. Then there exists a field $\mathfrak{S}$ which is an extension of $\mathfrak{F}$ such that $\mathcal{A}$ over $\mathfrak{S}$ is a total matrix algebra.*

Proof. Let $\mathcal{A} = \mathcal{M} \otimes \mathcal{D}$ where $\mathcal{M}$ is a total matrix algebra and $\mathcal{D}$ a division algebra over $\mathfrak{F}$. Then by Corollary 11 we see that $\mathcal{A}$ is a central simple algebra over the field $\mathfrak{D}$ which is the center of $\mathcal{D}$. Hence, by Theorem 17, there exists a scalar extension $\mathfrak{S}$ of $\mathfrak{D}$ such that $\mathcal{A}$ over $\mathfrak{S}$ is a total matrix algebra. Clearly, $\mathfrak{S}$ is an extension of $\mathfrak{F}$.

Theorem 18 shows how closely a simple algebra resembles a total matrix algebra. Lemma 14 and Theorem 19 below indicate another instance of their resemblance.

LEMMA 14. *Let $\mathcal{M}$ be a total matrix algebra. Then $\mathcal{M}$ has no basis every element of which is nilpotent.*

Proof. Without loss of generality, we suppose that $\mathcal{M}$ is a 2 by 2 total matrix algebra over a field $\mathfrak{F}$ and that $\{B_1, B_2, B_3, B_4\}$ is a basis of $\mathcal{M}$. Assume the contrary that each B_i is a 2 by 2 nilpotent matrix. But then since $\begin{pmatrix} 1 & 0 \\ 0 & 0 \end{pmatrix}$ is an element of $\mathcal{M}$ we must have

$$\begin{pmatrix} 1 & 0 \\ 0 & 0 \end{pmatrix} = \sum_{i=1}^{4} s_i B_i \qquad \text{for some} \qquad s_i \in \mathfrak{F}. \tag{72}$$

However, in view of Lemma 18 on page 98 and the fact that the traces of similar matrices are equal, it follows from (72) that

$$\text{tr} \begin{pmatrix} 1 & 0 \\ 0 & 0 \end{pmatrix} = 1 = \text{tr} \sum_{i=1}^{4} s_i B_i = \sum_{i=1}^{4} s_i (\text{tr } B_i) = 0$$

which is a contradiction. Hence our assumption is false and the Lemma is proved.

THEOREM 19. *Let $\mathscr{A}$ be a simple algebra. Then $\mathscr{A}$ has no basis every element of which is nilpotent.*

Proof. Let $\mathscr{A}$ be a simple algebra over a field $\mathfrak{F}$ and $\{B_1, \ldots, B_n\}$ be a basis of $\mathscr{A}$. Assume the contrary that each B_i is nilpotent. But then by Theorem 18 we see that $\mathscr{A}$ is a total matrix algebra over a field $\mathfrak{S}$ which is an extension of $\mathfrak{F}$. However, a subset of $\{B_1, \ldots, B_n\}$ can serve as a basis of the total matrix algebra $\mathscr{A}$ over $\mathfrak{S}$ which contradicts Lemma 14. Hence our assumption is false and the Theorem is proved.

Based on Theorem 19 we prove

THEOREM 20. *An algebra is nilpotent if and only if it has a basis every element of which is nilpotent.*

Proof. If $\mathscr{A}$ is a nilpotent algebra then clearly every element of $\mathscr{A}$ is nilpotent and hence every element of any basis of $\mathscr{A}$ is nilpotent.

Next, let $\mathscr{A}$ be an algebra with a basis every element of which is nilpotent. Assume the contrary that $\mathscr{A}$ is not nilpotent and let $\mathscr{N}$ be the radical of $\mathscr{A}$. Thus, by our assumption $\mathscr{A} \neq \mathscr{N}$. By Theorem 8 on page 145 we see that $\mathscr{A}/\mathscr{N}$ is semi-simple and since $\mathscr{A}/\mathscr{N}$ is homomorphic to $\mathscr{A}$ it has a basis every element of which is nilpotent. But then from Theorem 12 on page 146 it follows that $\mathscr{A}/\mathscr{N}$ is a simple algebra or is a direct sum of simple algebras where each summand is homomorphic to $\mathscr{A}/\mathscr{N}$. However, since $\mathscr{A}/\mathscr{N}$ has a basis every element of which is nilpotent we see that our assumption implies the existence of a simple algebra with a basis every element of which is nilpotent. Clearly, this contradicts Theorem 19. Hence our assumption is false and the Theorem is proved.

Exercises

1. Using Lemma 13 prove that $\mathscr{A} = \mathscr{B} \otimes \mathscr{C}$ is a central algebra if and only if $\mathscr{B}$ and $\mathscr{C}$ are central algebras. Moreover, $\mathscr{B}$ is the centralizer of $\mathscr{C}$ in $\mathscr{A}$ and $\mathscr{C}$ is the centralizer of $\mathscr{B}$ in $\mathscr{A}$.

2. Let $\mathscr{A} = \mathscr{B} \otimes \mathscr{C}$ be a simple algebra. Prove that $\mathscr{B}$ as well as $\mathscr{C}$ is simple.

3. Based on Definition 10 on page 107, prove that if $\mathscr{A}$ is a central simple algebra and $\mathscr{A}'$ is the reciprocal algebra of $\mathscr{A}$ then $\mathscr{A} \otimes \mathscr{A}'$ is a total matrix algebra.

4. Prove that an algebra $\mathscr{A}$ over a field $\mathfrak{F}$ is central simple if and only if $\mathscr{A}$ over $\mathfrak{S}$ is central simple for every scalar extension $\mathfrak{S}$ of $\mathfrak{F}$.

5. Let I be the unity of a division algebra $\mathscr{D}$ of dimension greater than one over a field $\mathfrak{F}$. Let D be an element of $\mathscr{D}$ such that $D = sI$ for no element s of $\mathfrak{F}$. Let $\mathfrak{S}$ be the field obtained by adjoining to $\mathfrak{F}$ a root of the minimum polynomial of D. Prove that $\mathscr{D}$ over $\mathfrak{S}$ is not a division algebra.

6. Prove Theorem 17 based on Theorem 15 and Problems 4 and 5.

7. Prove that the dimension of a central simple algebra is equal to n^2 where n is a natural number.

8. Prove that the dimension of a simple algebra is equal to m^2n where m and n are natural numbers.

9. Prove that a simple algebra over the field of real numbers is either a total matrix algebra or a tensor product of a total matrix algebra and the algebra of ordinary complex numbers or a tensor product of a total matrix algebra and the algebra of real quaternions.

4.5. Concluding Remarks

Based on the results of the previous sections of this Chapter, we may classify algebras as follows.

An algebra is either nilpotent or nonnilpotent. In Section 4.2 some properties of a nilpotent algebra (which clearly is equal to its radical) were established. A nonnilpotent algebra is either without radical or with radical. A nonnilpotent algebra without radical is called a semi-simple algebra. In Section 4.3 it was established that a semi-simple algebra is either a direct sum of simple algebras or is a simple algebra. In Section 4.4 it was established that a simple algebra is a tensor product of a total matrix algebra and a division algebra.

We have not considered the case of a nonnilpotent algebra with radical. A main result pertaining to this case is the so-called *Principal Theorem of Wedderburn* (*see* the theorem below) which involves the notion of a separable algebra.

Clearly, if an algebra $\mathscr{A}$ over a field $\mathfrak{F}$ is semi-simple over a scalar extension field of $\mathfrak{F}$ then $\mathscr{A}$ is semi-simple over $\mathfrak{F}$. However, a semi-simple algebra over a field $\mathfrak{F}$ need not be a semi-simple over every scalar extension field of $\mathfrak{F}$. If an algebra $\mathscr{A}$ over a field $\mathfrak{F}$ is semi-simple over every scalar extension field of $\mathfrak{F}$ then $\mathscr{A}$ is called a *seperable algebra*.

Now, let $\mathscr{A}$ be a nonnilpotent algebra over a field $\mathfrak{F}$ and let $\mathscr{N} \neq \{0\}$ be the radical of $\mathscr{A}$. Clearly, $\mathscr{N}$ is an ideal of $\mathscr{A}$ and $\mathscr{A}/\mathscr{N}$ is a semi-simple algebra (*see* Theorem 8 on page 145). But since $\mathscr{N}$ has no unit element, $\mathscr{N}$ is not a direct summand of $\mathscr{A}$, in general (*see* Theorem 1 on page 125). Moreover, in general, $\mathscr{N}$ is not even a supplementary summand of $\mathscr{A}$, in the sense of $\mathscr{I}$ in Lemma 12 on page 87. However, if the semi-simple algebra $\mathscr{A}/\mathscr{N}$ over $\mathfrak{F}$ is separable, then (according to the theorem below) $\mathscr{N}$ is a supplementary summand of $\mathscr{A}$ to which Lemma 12 on page 87 applies.

THEOREM. *Let $\mathscr{A}$ be an algebra with radical $\mathscr{N}$ such that $\mathscr{A}/\mathscr{N}$ is separable. Then*

$$\mathscr{A} = \mathscr{B} \dotplus \mathscr{N}$$

where $\mathscr{B}$ is a subalgebra of $\mathscr{A}$ isomorphic to $\mathscr{A}/\mathscr{N}$.

Clearly, the above theorem is of particular interest for the case where $\mathscr{A}$ is nonnilpotent and $\mathscr{N} \neq \{0\}$. We omit the proof of the above theorem and mention only that its proof is partly based on Lemma 12 on page 87, Lemma 13 on page 88, Theorem 8 on page 145, Theorem 12 on page 146, Theorem 15 on page 148 and Theorem 18 on page 154.

Bibliography

Albert, A. A., *Modern Higher Algebra*. Chicago: University of Chicago Press, 1947.

Albert, A. A., *Structure of Algebras*. New York: American Mathematical Society, 1939.

Artin, E., Nesbitt, C. J. and Thrall, R. M., *Rings with Minimum Condition*. Ann Arbor: University of Michigan Press, 1944.

Barnes, W. E., *Introduction to Abstract Algebra*. Boston: Heath, 1967.

Birkhoff, G. and MacLane, S., *Algebra*. New York: Macmillan, 1967.

Chevalley, C., *Fundamental Concepts of Algebra*. New York: Academic Press, 1956.

Clifford, A. H. and Preston, G. B., *The Algebraic Theory of Semigroups*, 2 vols. Providence, R.I.: American Mathematical Society Surveys No. 7, 1961.

Curtis, C. W. and Reiner, I., *Representation Theory of Finite Groups and Associative Algebras*. New York: Interscience, 1966.

Dickson, L. E., *Algebras and their Arithmetic*. New York: Dover, 1960.

Halmos, P. R., *Finite Dimensional Vector Spaces*. Princeton: Van Nostrand, 1958.

Herstein, I. N., *Topics in Algebra*. New York: Blaisdell, 1964.

Jacobson, N., *Lectures in Abstract Algebra*, 3 vols. Princeton: Van Nostrand, 1953.

Jacobson, N., *Structure of Rings*. Providence, R. I.: American Mathematical Society, 1956.

Jacobson, N., *The Theory of Rings*. New York: American Mathematical Society, 1943.

Kurosh, A. G., *Lectures on General Algebra*. New York: Chelsea, 1963.

Lang, S., *Algebra*. Reading Mass.: Addison-Wesley, 1967.

MacDuffee, C. C., *An Introduction to Abstract Algebra*. New York: Wiley, 1940.

McCoy, N. H., *Rings and Ideals*. Buffalo: The Mathematical Association of America, 1948.

McCoy, N. H., *The Theory of Rings*. New York: Macmillan, 1964.

Parker, W. V. and Eaves, J. C., *Matrices*. New York: Ronald Press, 1960.

Peirce, B., Linear Associative Algebra. *American Journal of Mathematics*, **6** (1881), 97.

Perlis, S., *Theory of Matrices*. Reading, Mass.: Addison-Wesley, 1958.

Van der Waerden, B. L., *Modern Algebra*, 2 vols. New York: Frederick Ungar, 1950, 1953.

Vinograde, B., *Linear and Matrix Algebra*. Boston: Heath, 1967.

Wedderburn, J. H. M., *Lectures on Matrices*. New York: American Mathematical Society, 1934.

Wedderburn, J. H. M., *On Hypercomplex Numbers*. Proceedings of London Mathematical Society (2)**6**, (1908), 77–118.

Zariski, O. and Samuel, P., *Commutative Algebra*, 2 vols. Princeton: Van Nostrand, 1958, 1960.

Index of Symbols

Index